Organic Chemistry 1

Learning With AI Tools

Resource for Peer-Led Team Learning Workshop

Ana Fraiman, Ph.D., Professor, Audrey L. Reynolds Distinguished Teaching Professor Department of Chemistry, Northeastern Illinois University

Emily Fioramonti, Peer Leader, Department of Biology
Northeastern Illinois University

Paras Mehta, Peer Leader, Department of Chemistry
Northeastern Illinois University

Alexandra Gokee, Peer Leader, Department of Chemistry
Northeastern Illinois University

Brenna Dooley, Peer Leader, Department of Biology
Northeastern Illinois University

PEER-LED TEAM LEARNING
INTERNATIONAL SOCIETY

Published by the Peer-Led Team Learning International Society

A non-profit organization for the advancement of learning

Organic Chemistry 1 Learning With AI Tools
Resource for Peer-Led Team Learning Workshop

Cover Art: Emily Fioramonti

10 9 8 7 6 5 4 3 2 1

ISBN 978-1-944996-03-1

Peer-Led Team Learning International Society (www.pltlis.org)

7100 Westwind Drive, Suite 240, El Paso, Texas 79912 USA
info@pltlis.org

Series Editors: James E. Becvar, Ph.D.
A.E. Dreyfuss, Ed.D.
Ana Fraiman, Ph.D.

Preface

Organic Chemistry workshop is based on Peer-Led Team Learning (PLTL), a method for improving learning outcomes for students through its emphasis on team-based strategies. PLTL has enabled thousands of students to learn course material since its inception in the 1990s, and its effectiveness is supported by extensive research. Originally with support from grants from the National Science Foundation and other funding sources, many variations of PLTL have evolved throughout the United States and globally.

The PLTL program provides students with many valuable tools for success. Some of the most important of these tools include:

- How to approach learning the practice of organic chemistry
- How to implement basic concepts of general chemistry into organic chemistry
- How to ask questions in order to solve problems
- How to cooperate with, work with, and learn from your peers and develop a network of reliable colleagues who can positively reinforce your learning environment
- How to reinforce your knowledge for examinations and beyond
- How to study effectively and efficiently
- Comparing and analyzing the information provided by AI tools.

Learning these skills is essential for students in every course and especially for those who have recently entered a university environment.

Students enrolled in an Organic Chemistry lecture course with PLTL are also expected to attend a separate, weekly workshop session. The workshops are composed of small groups of students from the larger lecture sections, who will work together on workshop activities for the entire semester. Workshop sessions include various problem-solving methods and activities, which help to cement knowledge, foment scientific curiosity, and demonstrate analytic approaches through real-world, individual and small group experiences.

The use of this resource book within the workshop environment facilitates the implementation of workshop activities and reinforces essential learning modes. Your peer leader will apply these materials as assignments and as in-workshop activities throughout the semester.

To the Student

Welcome to Organic Chemistry. The workshop part of this course provides a small group setting that will complement the larger lecture setting. Your workshop will be led by a peer leader, an undergraduate student who has demonstrated a strong command of Organic Chemistry and who is trained for this leadership position. This resource book helps you and the workshop peer leaders conduct each week's activities. Both the lecture and the workshop are guided by your professor; the workshop is led by the peer leader. Workshops meet at a separate time and location from the lecture. Workshops are designed to reinforce lecture topics through practice, discussion, and small group activities.

Strategies for Success

The most important thing you can do to ensure your success in Organic Chemistry is to be prepared for each lecture and each workshop. Good preparation entails keeping up to date with your homework assignments and reading the textbook sections that apply to the lectures. To truly excel, read about the Organic Chemistry topics before they are introduced in lecture. Good preparation will enable you to earn an "A" in this course with less time and effort.

There may be times you fall behind. When this happens, schedule extra time with a tutor or professor during their office hours or make appointments. This is the most effective way to catch up. By not fixing these predicaments as soon as they occur harms your ability to earn a good grade in the course. An essential skill in learning is admitting when you need help. Please take advantage of the resources that are provided for you as part of this course.

Keep your focus. If you have motivating factors behind your academic career, keep those factors in mind as you work hard. Those motivating factors are the key to perseverance. You may still be trying to discover what motivates you. For this, you will need to keep your eyes and mind open to the new learning experiences that a college education can offer.

Ana Fraiman, Ph.D.
Audrey L. Reynolds Distinguished Teaching Professor
Department of Chemistry, Northeastern Illinois University

~ The Series Editors

WELCOME

From the Peer Leaders to the Student:

Many students see organic chemistry as one of the most challenging courses in preparation for graduate and professional school exams. Organic chemistry requires a deep understanding of the core concepts, some memorization, and lots of practice applying those concepts. This workbook was written to be used as a supplemental guidebook with problem sets for additional practice, useful tips, and other helpful resources. The concepts covered in this workbook are the ones that we, as students, found to be more challenging, and crucial for understanding organic chemistry.

Each workshop starts with concept checks that contain a list of topics that we recommend reviewing in preparation for solving problems in that workshop. In concept checks we tried to include important rules, and principles, and also tools and guidelines that we found to be helpful. Make sure to read the concept checks thoroughly before attempting to solve any problems. Note that this workbook is not to be used to replace the textbook. Make sure to read assigned chapters and do the additional problems the textbook provides.

Table of Contents

WORKSHOP 1 Introduction and Review

Chemical Bonding, Hybridization, Acid-Base Properties

Concept Check:
- ✓ **Bonding types**
- ✓ **Acid / Base**
- ✓ **Molecular geometry & Hybridization**
- ✓ **Formal Charges**
- ✓ **Organic Chemistry conventions**
- ✓ **Use of AI**

Chemical Bond Types

Covalent:

σ: A single bond between two atoms in an organic molecule is referred to as a sigma bond. This involves an overlap of molecular orbitals, increasing electron density in the region between two atoms. This can also be thought of as a sharing between two electrons.

π: A parallel overlap of p orbitals above and below the linear nuclear interaction. This is presented as double or triple bonds.

Ionic:

Electrostatic interaction between cations and anions, generally metals and nonmetals.

Molecular Geometry

An explanation to experimentally obtained bond angles is a formation of hybrid orbitals between *s* and *p* orbitals to keep repelling pairs of electrons as far as possible.

For molecular geometries, refer to the inside front cover.

Acid Strength

The larger the dissociation constant (K_a), the lower the pKa, and the more likely the hydrogen will dissociate. This is used to determine acid strength.

$$HA \rightleftarrows H^+ + A^-$$

$$Henderson - Hasselbalch:$$

$$pH = pK_a + log\frac{[A^-]}{[HA]}$$

$$pK_a = -\log K_a$$

$$K_a = \frac{[H^+][A^-]}{[HA]}$$

INSTRUCTIONS FOR PROBLEM-SOLVING

1. *Solve each problem with your peers*
2. *Compare your answers*
3. *Check your answer with an AI model of your choice (use different AI model in your team)*
    a. *Was the answer the same or different than you and your peers obtained with different AI models?*
    b. *If the same: What additional information did you receive and from which AI model?*
    c. *If the answer is different: Why is the AI solution incorrect and where is the mistake?*
4. *Did you change your answer and why?*

WORKSHOP PROBLEMS

1. Draw the Lewis structures for each of the following compounds:

- Show all bonding and nonbonding valence electrons.
- Indicate the formal charges on atoms.
- Indicate the kind of bonding present: σ and π bonding.
- Describe the bonds and explain how the overall molecule is polar or nonpolar.
- Describe the geometry and hybridization of the bonding: sp3, sp2, or sp.

a) NaBr

b) Hydrobromic acid, hydrochloric acid, hydroiodic acid, hydrofluoric acid

(Indicate different acid strengths, explain)

c) CH_3CH_3, PCl_3, acetone, C_3H_7NO

d) Carboxylic acid, carboxylate ion

e) Dimethyl sulfoxide

2. We accept the idea that the hybridization of tetra-substituted carbon is sp3 and the geometry is tetrahedral:

How many isomers are possible for CH_4, CH_3I, CH_2Br_2, CH_2BrIO?

Work with models to solve the problem.

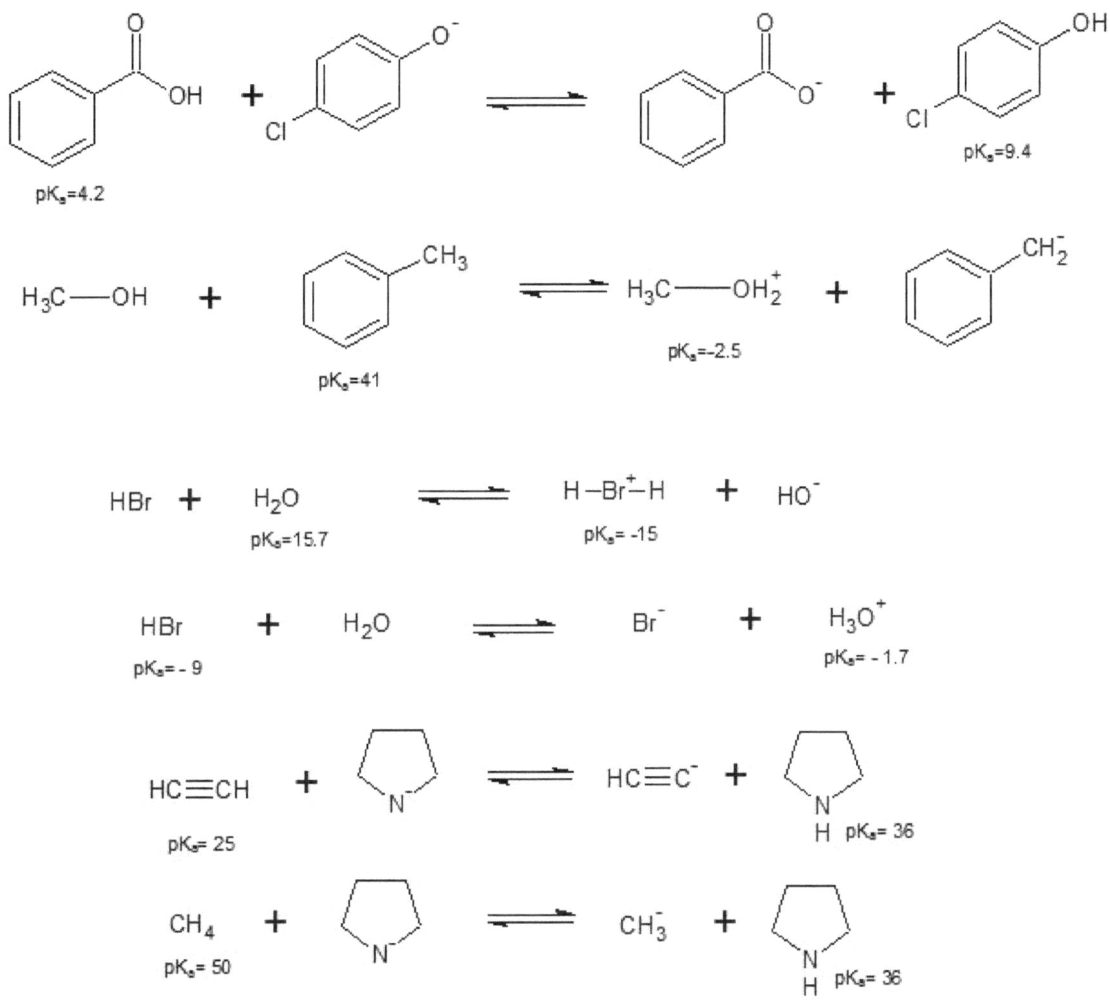

3. For each of the following reactions:

- Draw complete Lewis dot structures,
- Show all bonding and nonbonding electron pairs,
- Use curved arrows to show the flow of electrons as the reaction proceeds.
- Identify the acids, show the major species at equilibrium.

4. For each of the following pairs, one reaction proceeds to the specified products and the other does not. Predict which is which and clearly explain your choice. It will help to start with curved arrows to show which bonds are made and which bonds are broken in each reaction.

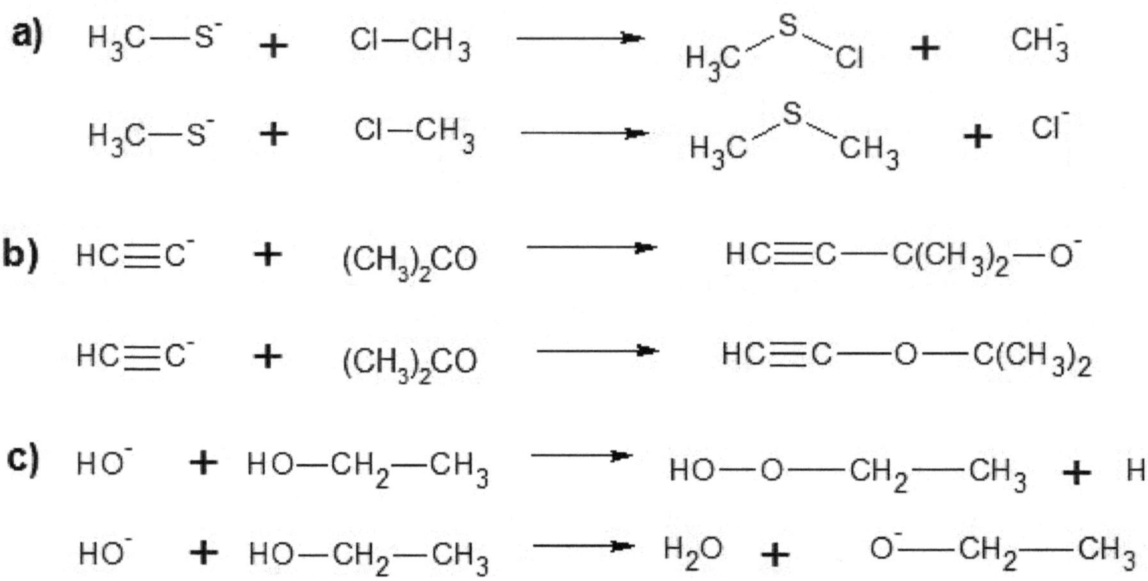

5. Answer the following for each given molecule:

a. How many carbons are in the following drawing? How many hydrogens?

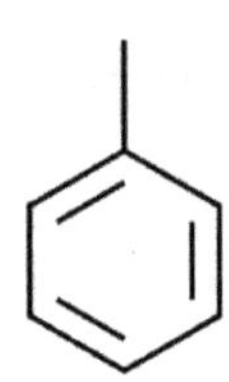

b. How many carbons are in the following drawing? How many hydrogens?

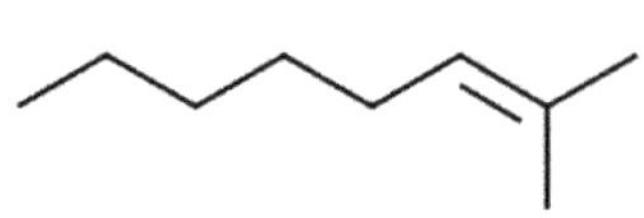

c. How many carbons are in the following drawing? How many hydrogens?

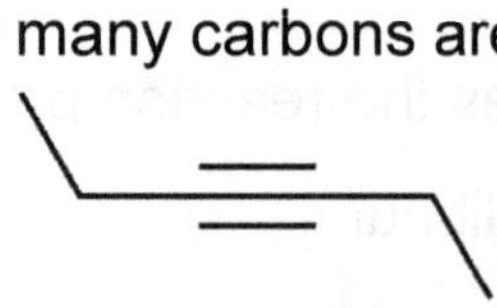

d. Look at the following molecule and draw in the hydrogens and electron pairs:

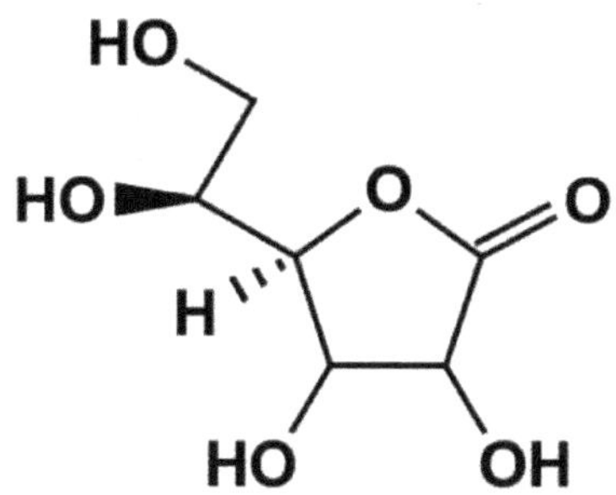

(hint: Do all of the carbons have 4 bonds? Do all of the oxygens have a full octet?)

6. Answer the following questions, fill in the blank, and complete the structure:

a) How many bonds can hydrogen make?

b) How many bonds can chlorine make?

c) Dashed lines show atomic bond that goes ____________(away or toward) you.

d) Draw $ClCH_2CH_2CH(OCH_3)CH_3$ in Lewis dot and line-angle forms.

WORKSHOP 2 Structure of Alkanes. Alkyl Substitution

Concept Check:
- ✓ **Types of alkyl substitution**
- ✓ **Degrees of unsaturation**
- ✓ **Radical halogenation**

Types of Carbon and Hydrogen

Degrees of Alkyl Substitution of Carbons			
Primary	**Secondary**	**Tertiary**	**Quaternary**
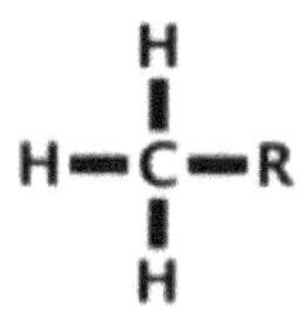	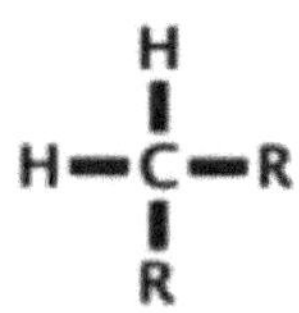	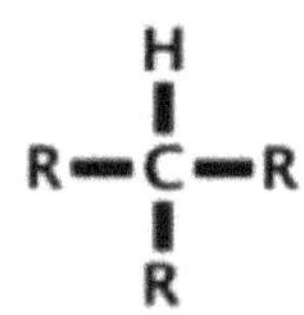	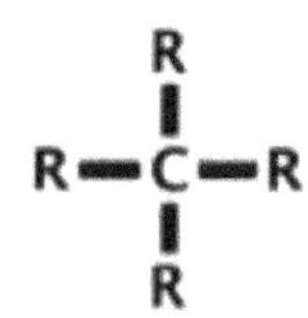

*Degree of substitution on carbon determines the type of hydrogen it is attached to.

Dr. Fraiman's Method for Calculating Degrees of Unsaturation

1. Draw the structure of the molecule with only one carbon and the heteroatoms given in the formula (the molecule does not have to exist, but it has to be fully saturated).

2. Write a general formula for the molecule you drew that is always expressed as $C_nH_{2n+/-x}$. This is the formula for the most saturated compound.

3. Apply the general formula to find the formula for the most saturated compound. Subtract the molecular formula given to you from the most saturated formula that you created. This gives you the amount of hydrogens missing. Divide the result by 2, this is the degree of unsaturation for this molecule.

Example:

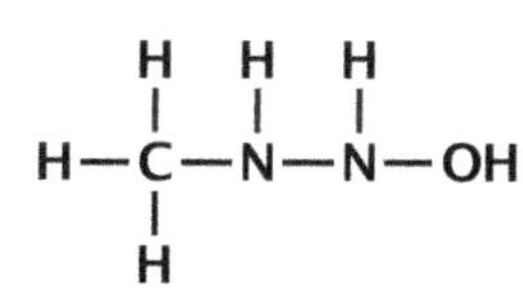

$C_5H_{10}N_2O$ is the molecular formula.

General formula for the drawn molecule: $C_nH_{2n+4}N_2O$

Formula for the most saturated compound: $C_5H_{14}N_2O$

Apply the formulas to calculate degree of unsaturation:

$$\begin{array}{r} - \; C_5H_{14}N_2O \\ \underline{C_5H_{10}N_2O} \\ 4/2 = 2 \end{array}$$

Radical Halogenation

Monohalogenation

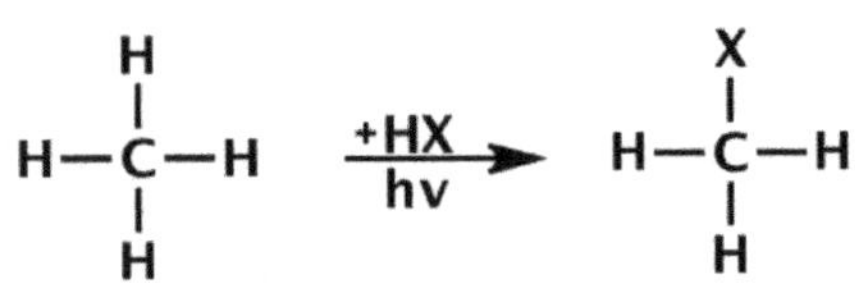

Dihalogenation

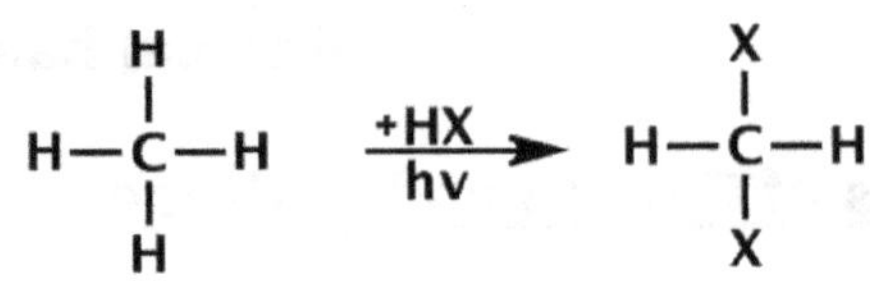

INSTRUCTIONS FOR PROBLEM-SOLVING

1. *Solve each problem with your peers*
2. *Compare your answers*
3. *Check your answer with an AI model of your choice (use different AI model in your team)*
    a. *Was the answer the same or different than you and your peers obtained with different AI model?*
    b. *If the same: What additional information did you receive and from which AI model?*
    c. *If the answer is different: Why is the AI solution incorrect and where is the mistake?*
4. *Did you change your answer and why?*

WORKSHOP PROBLEMS

1. The following molecular formula C_8H_{18} can have many different isomers.

a) Draw a structural isomer with only primary and secondary carbons.

b) Draw a structural isomer with primary, secondary, tertiary, and quaternary carbons.

c) Draw a structural isomer with primary, secondary and tertiary carbons.

2. Draw and name all possible isomers of C_8H_{18} and identify different kinds of carbons.

3. The following isomers of molecular formula C_5H_{12} underwent a chemical reaction to give all the monochloro derivatives (one H atom was substituted by a Cl atom).

$$C_5H_{12} \xrightarrow[\text{Radiation}]{Cl_2} C_5H_{11}Cl + HCl$$

a) Draw a C_5H_{12} compound that gives three $C_5H_{11}Cl$ products.

b) Draw a C_5H_{12} compound that gives four $C_5H_{11}Cl$ products.

c) Draw a C_5H_{12} compound that gives only one $C_5H_{11}Cl$ product.

4. a) Draw 5 structural isomers of C_7H_{14}.

b) Draw all of the monobrominated products of those isomers.

5. Draw and name all of the dichlorinated products of butane.

WORKSHOP 3 Molecular Geometry: Isomerism

Concept Check:
- ✓ **Hybridization**
- ✓ **Molecular geometry**
- ✓ **Isomeric relationships**
- ✓ **Structural and geometric isomers**
- ✓ **Organic nomenclature**

Relationship of hybridization to bond angles and molecular geometries

# of orbitals	Hybridization	Angle	Geometries
2	sp	180	linear
3	sp2	120	trigonal planar
4	sp3	109.5	trigonal pyramidal, tetrahedral

For isomeric relationships, refer to the inside front cover.
For general naming rules, refer to the inside front cover.

Rules for naming complex substituents

1. Find main carbon chain (rings take priority).
2. Find longest chain in the complex substituent.
3. C1 of the complex substituent is directly attached to the main chain.
4. Name complex group according to IUPAC rules.
5. Parentheses are used to name complex group incorporated in the entire name.

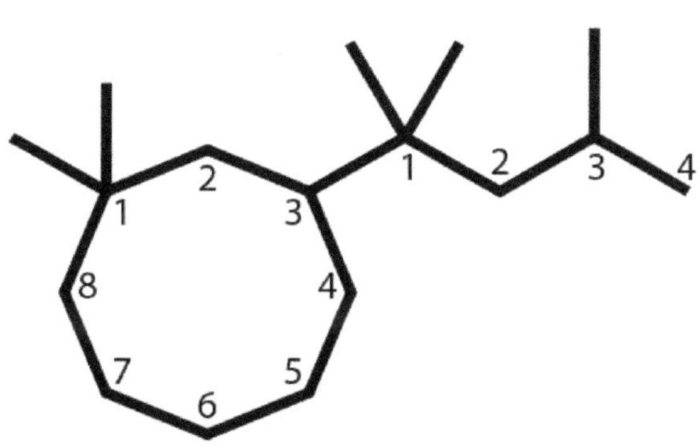

1,1-dimethyl-3-(1,1,3-trimethylbutyl)cyclooctane

INSTRUCTIONS FOR PROBLEM-SOLVING

1. *Solve each problem with your peers*
2. *Compare your answers*
3. *Check your answer with an AI model of your choice (use different AI model in your team)*
   a. *Was the answer the same or different than you and your peers obtained with different AI model?*
   b. *If the same: What additional information did you receive and from which AI model?*
   c. *If the answer is different: Why is the AI solution incorrect and where is the mistake?*
4. *Did you change your answer and why?*

WORKSHOP PROBLEMS

1. Predict the hybridization and geometry of every atom in the following ions:

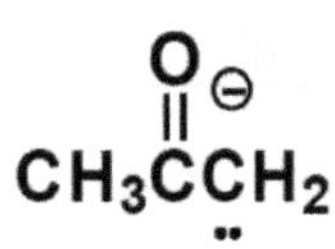

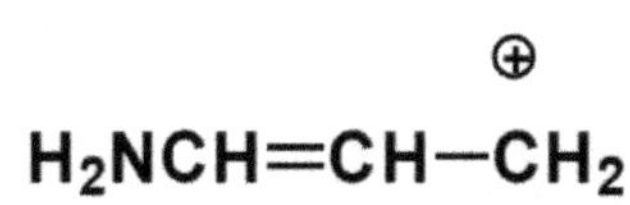

2. Determine the relationships between any and all pairs of the following compounds:
(Options: structural isomers, geometric isomers, identical, or not related)

a)
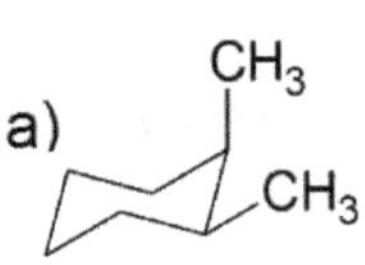

b)
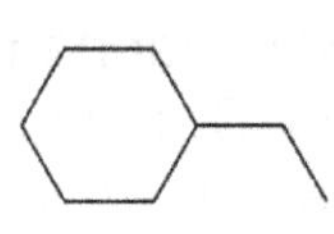

c)
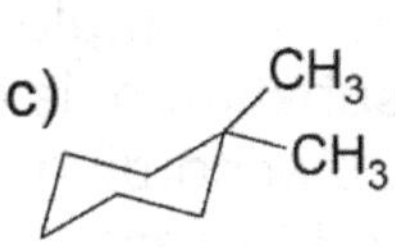

d)
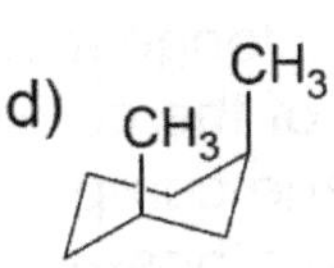

e)
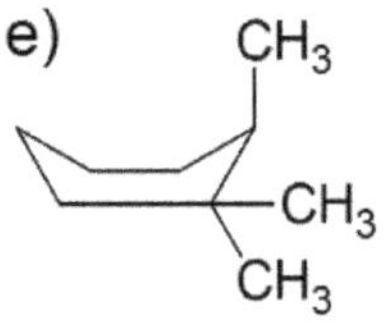

f)
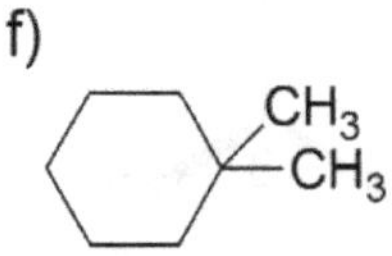

g)
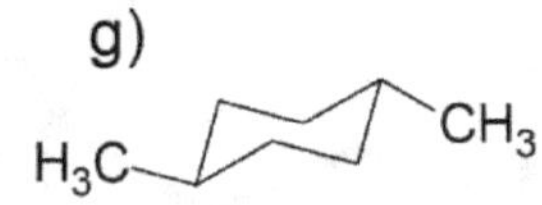

h)
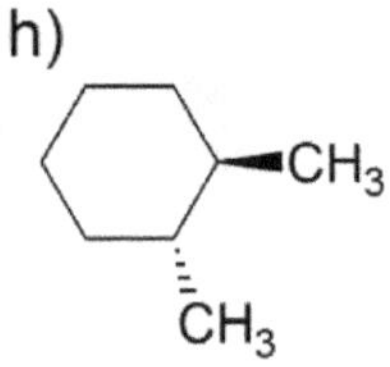

3. Give the IUPAC names for the following compounds:

a) 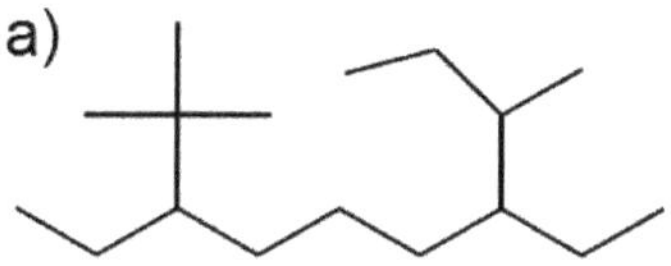

b) 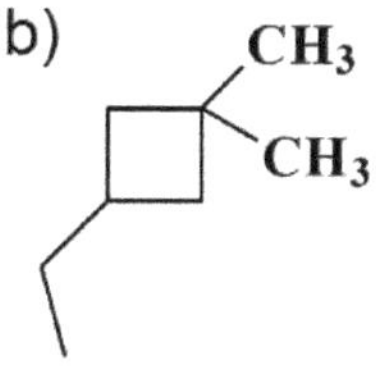

c) 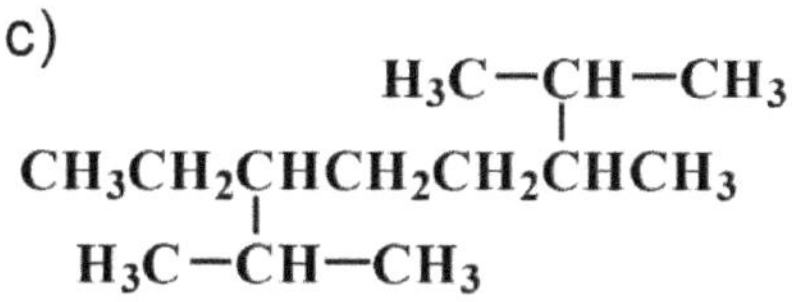

d) 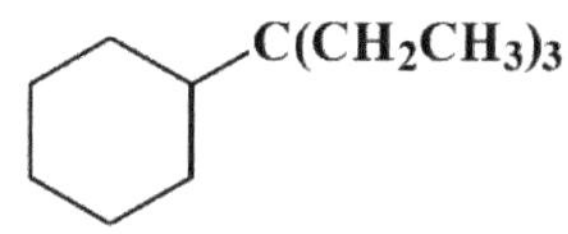

4. Draw and name eight isomers of molecules with the formula $C_8H_{17}Cl$ and $C_7H_{13}Br$.

WORKSHOP 4 Conformations of Alkanes

Concept Check:

- ✓ **Newman projections**
- ✓ **Potential energy diagram**
- ✓ **Chair conformations**

Newman projections are a tool we use to visualize 3-dimensional molecules on a 2-dimensional plane, while preserving structural information about conformational stabilities. The projections focus on a single carbon-carbon bond in a molecule, showing the 3D arrangement of the groups attached to those carbons. The major conformations we will discuss are eclipsed, fully eclipsed (eclipsed with the largest groups overlapping), gauche, and anti.

fully eclipsed
(eclipsed)
$\theta_{Me/Me}= 0^{\circ}$

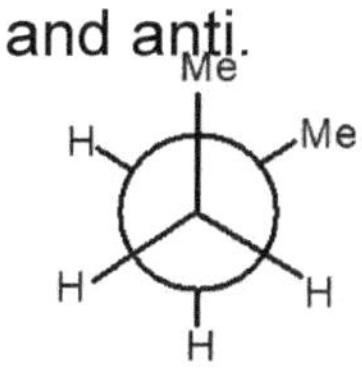

gauche
(staggered)
$\theta_{Me/Me}= 60^{\circ}$

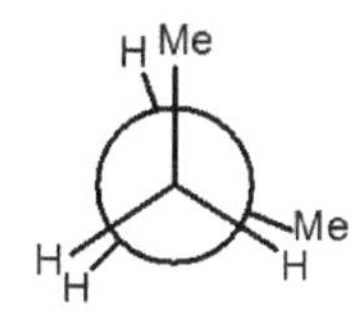

eclipsed
(eclipsed)
$\theta_{Me/Me}= 120^{\circ}$

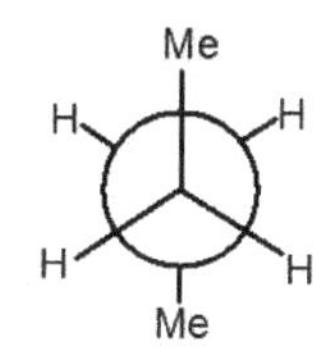

anti
(staggered)
$\theta_{Me/Me}= 180^{\circ}$

Newman Projections in Chairs

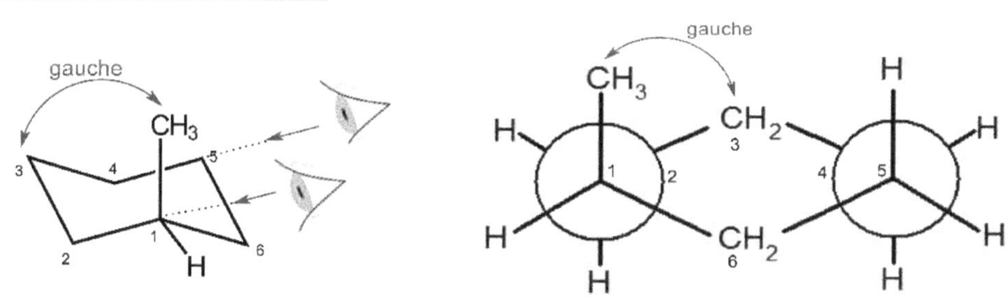

Newman projection
(a)

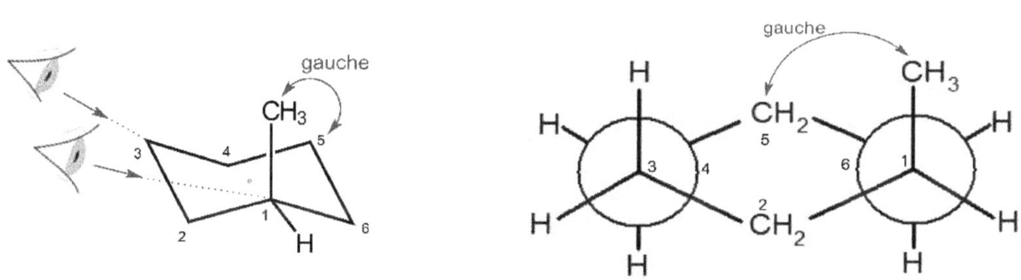

Newman projection
(b)

a) When the $-CH_3$ is in an axial position on C1, it is gauche to C3.
b) The axial $-CH_3$ on C1 is also gauche to C5 of the ring.

A **potential energy diagram** is a graph which shows the changes in potential energy of a molecule. The figure below sows how potential energy changes with the rotation around the sigma bond in between 2 carbons. The x-axis represents the change of dihedral angle as the bond rotates. The y-axis represents the potential energy associated with the conformational changes taking place during the rotation. By examining the steric interactions of groups on adjacent carbons, we can make predictions about the stability of conformations.

There is an infinite number of conformational isomers, however we only look at the conformations associated with the highest and the lowest potential energies. As an example, consider the potential energy diagram of *n*-butane for rotation around C2-C3. In a 360-degree rotation there are 3 higher and 3 lower energy levels associated with the particular eclipsed and staggered conformations that result from the rotation.

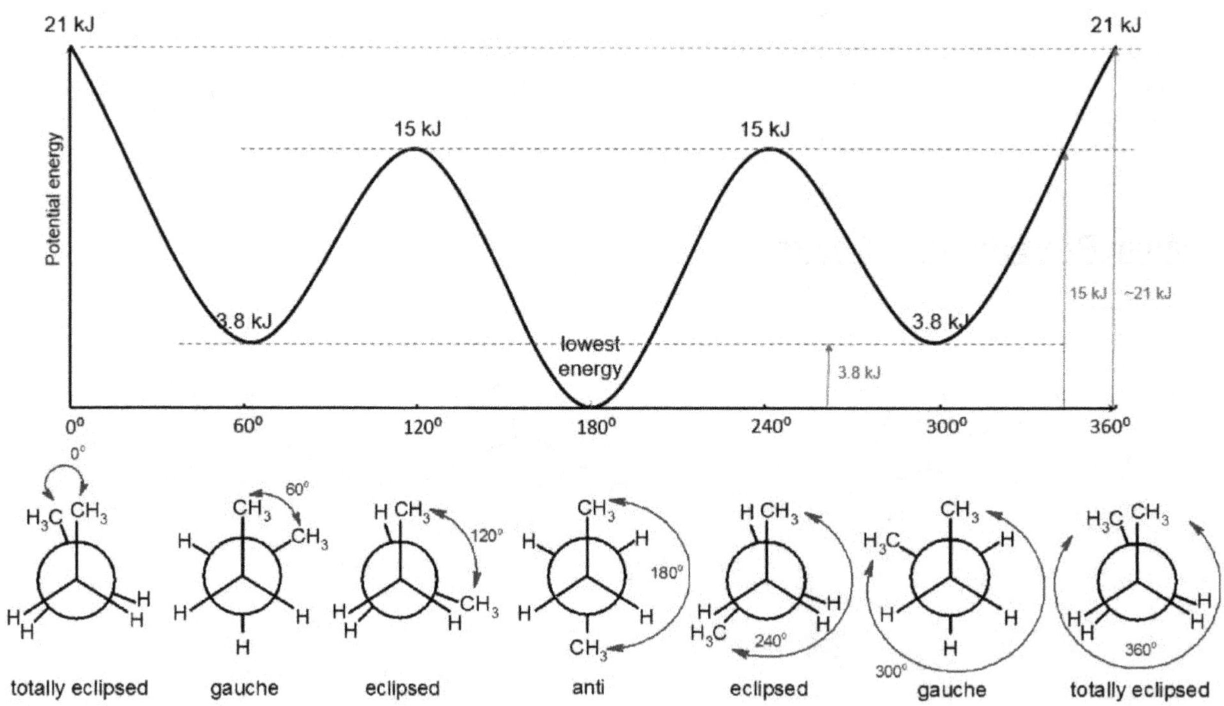

How to draw the perfect chair conformation

Step One

Step Two

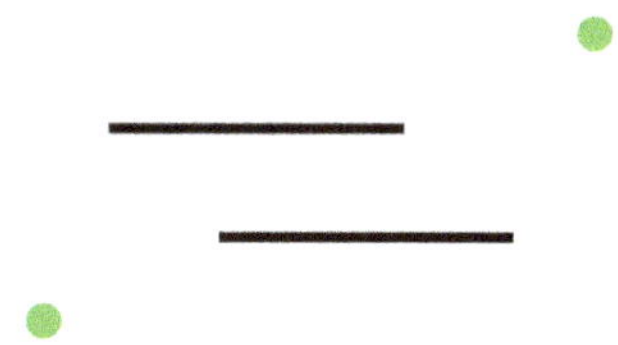

Step Three

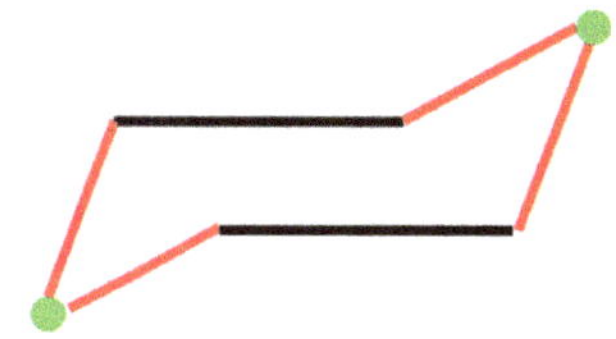

Recognizing Cis and Trans Isomers

To determine whether a chair confirmation of a di-substituted cyclohexane is a cis- or trans-isomer, you can convert the chair to a ring form. The axial and equatorial substituents alternate on the wedges and dashes, as seen below:

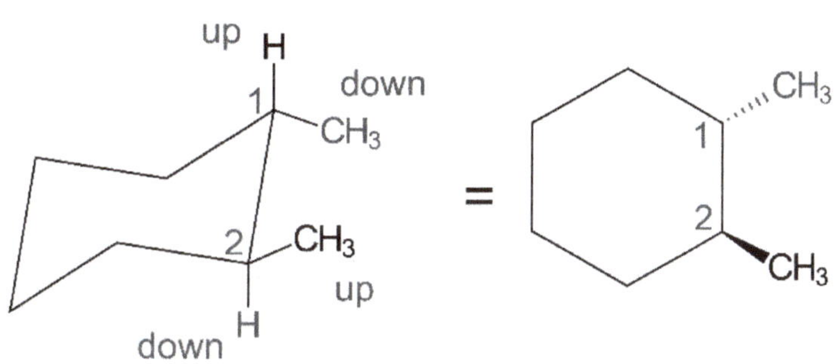

trans-1,2-dimethylcyclohexane

Interconversion of Chair Conformations

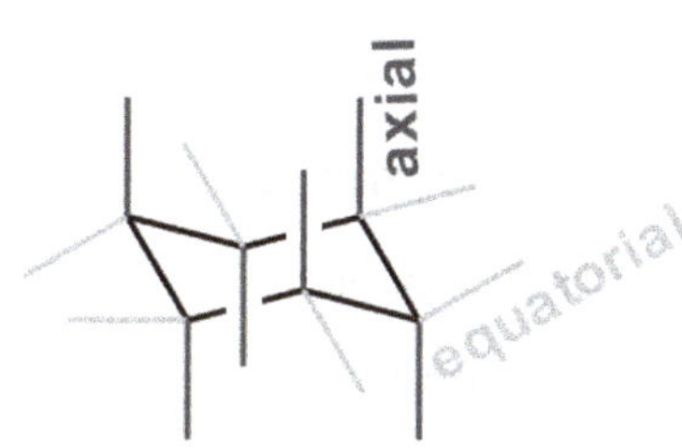

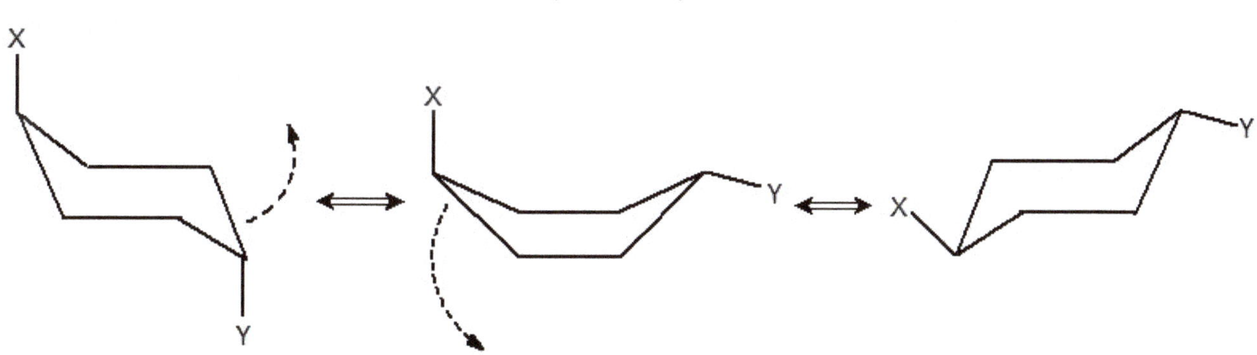

INSTRUCTIONS FOR PROBLEM-SOLVING

1. *Solve each problem with your peers*
2. *Compare your answers*
3. *Check your answer with an AI model of your choice (use different AI model in your team)*
    a. *Was the answer the same or different than you and your peers obtained with different AI model?*
    b. *If the same: What additional information did you receive and from which AI model?*
    c. *If the answer is different: Why is the AI solution incorrect and where is the mistake?*
4. *Did you change your answer and why?*

WORKSHOP PROBLEMS

1. Write Newman projections showing all possible staggered conformations about the C-2 C-3 bond in pentane.

a) Construct a potential energy diagram for rotations about the C-2-C-3 bond.

b) What is different and what is similar with the potential energy diagram for butane about the C-2-C-3 bond and the one you drew in part (a)?

c) What kind of generalization can be made regarding all *n*-chain alkanes?

2. **a)** Construct a potential energy diagram for 2,3-dimethyl butane.

b) Draw all the products of monobromination of 2,3-dimethyl butane.

c) Draw all the products of dibromination of 2,3-dimethyl butane.

3. **a)** Draw Newman projections for the C-1-C-2 bond of *cis*-1-bromo-2-methycyclohexane and *trans*-1-bromo-2-methycyclohexane in the most stable conformation and in the most unstable chair conformation.

b) Which one of the geometrical isomers is more stable? Explain the stability according to anti and gauche interactions and equatorial vs. axial substituents.

4. Which of the following isomers is most stable and why?

a)

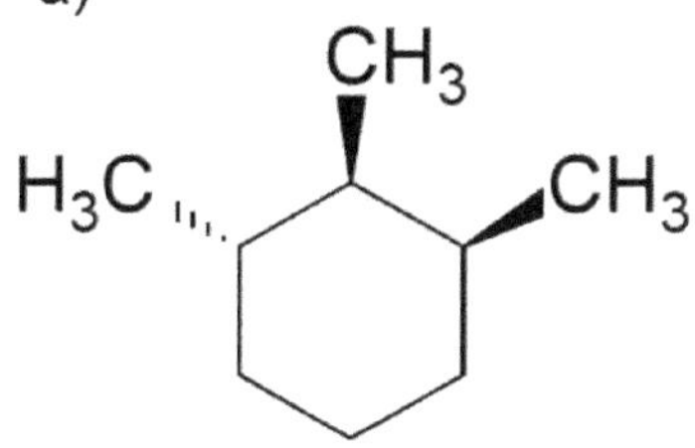

b)

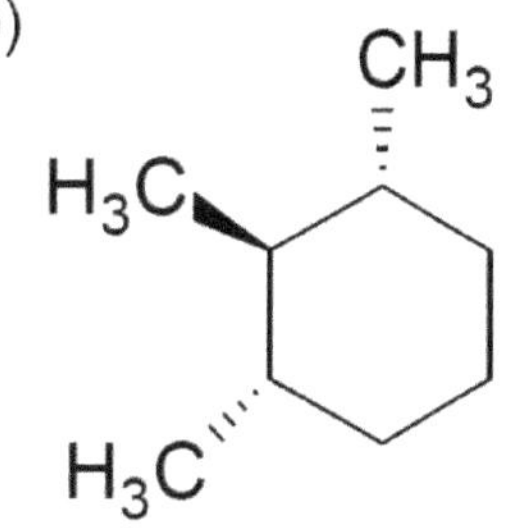

5. On viewing the drawings of *cis-trans*-decaline, which do you think is more stable and why?

a)

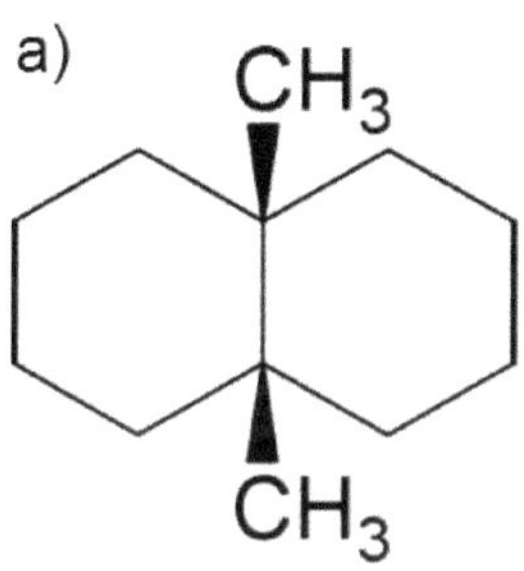

b)

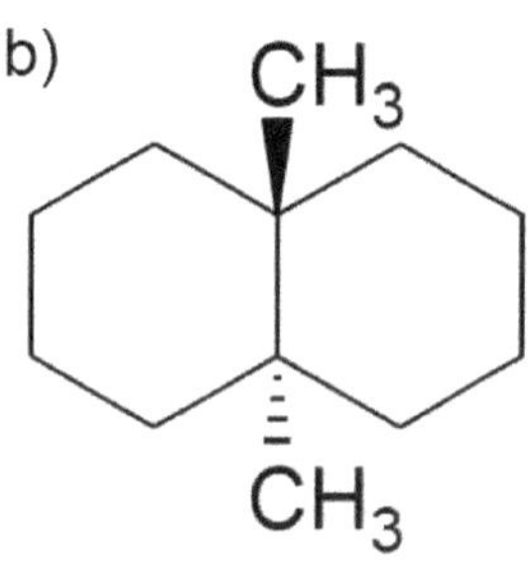

WORKSHOP 5 The Study of Chemical Reactions: Radical Reactions

Concept Check:

- ✓ **Mechanisms of reaction**
- ✓ **Free radical halogenation**
- ✓ **Major and minor products**
- ✓ **Bond dissociation enthalpies**

***Mechanism*: Free Radical Halogenation**

Initiation

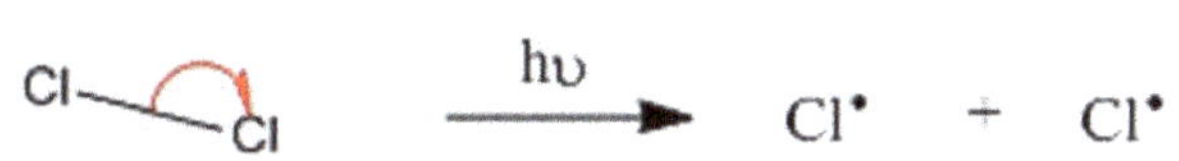

Propagation

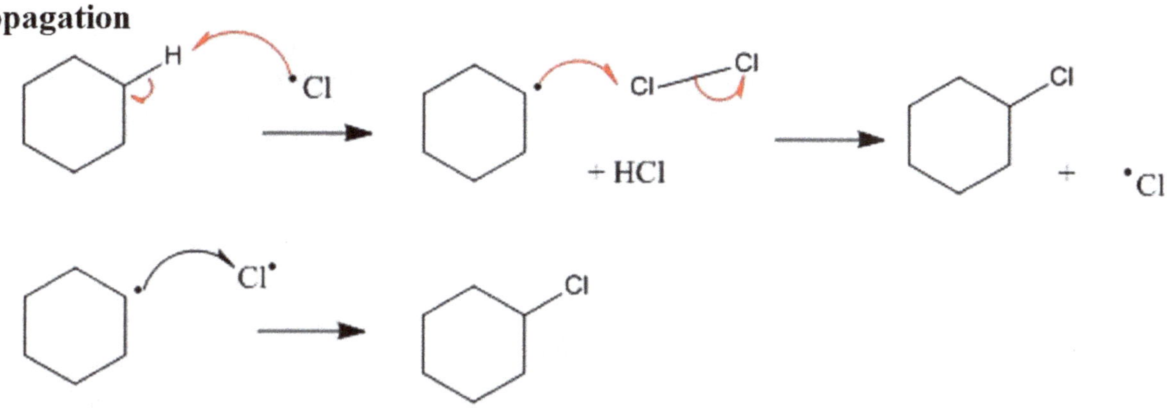

Termination

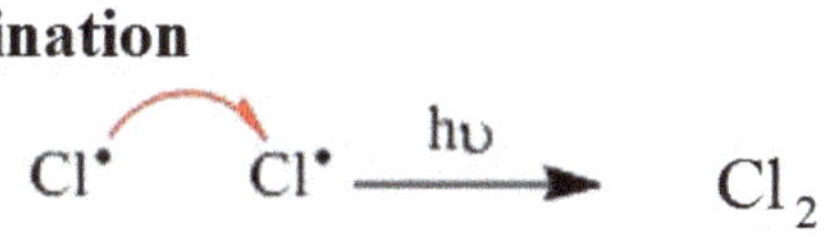

Homolytic vs. Heterolytic Cleavage

Homolytic cleavage (forms free radicals)

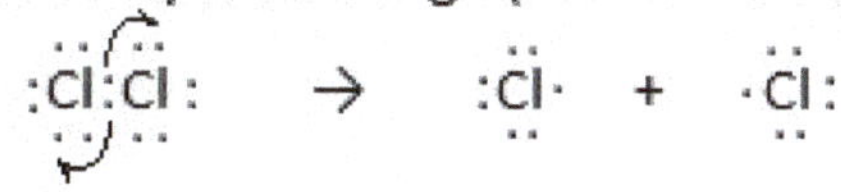

Heterolytic cleavage (forms ions)

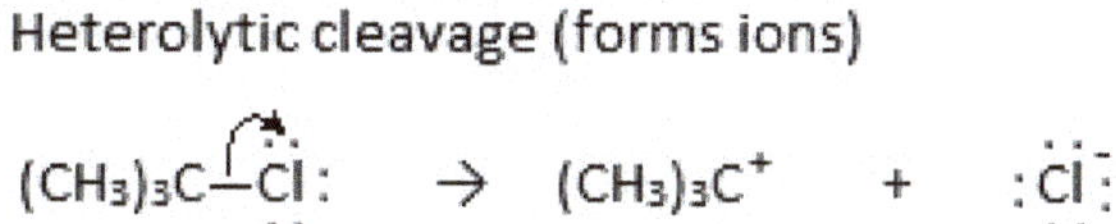

Bond Dissociation

$$\Delta H^\circ = \sum(BDE\ of\ bonds\ broken) - \sum(BDE\ of\ bonds\ formed)$$

Bonds to 1° carbons		Bonds to 2° carbons		Bonds to 3° carbons		H-X and X-X bonds	
Bond	kJ/mol	Bond	kJ/mol	Bond	kJ/mol	Bond	kJ/mol
RCH_2—H	410	$(R)_2CH$—H	397	$(R)_3C$—H	381	Cl—Cl	242
RCH_2—Cl	339	$(R)_2CH$—Cl	335	$(R)_3C$—Cl	331	Br—Br	192
RCH_2—Br	285	$(R)_2CH$—Br	285	$(R)_3C$—Br	272	H—Cl	431
						H—Br	368

Overall reaction

CH_3 – H + Cl – Cl → CH_3 – Cl + H – CL

Bonds broken ΔH° (per mole)		***Bonds formed ΔH° (per mole)***	
Cl—Cl	+242 kJ (+58 kcal)	H—Cl	-431 kJ (-103 kcal)
CH_3—H	+435 kJ (+104 kcal)	CH_3—Cl	-351 kJ (-84 kcal)
Total	+677 kJ (+162 kcal)	Total	-782 kJ (-187 kcal)

ΔH° = +677 kJ/mol + (-782) kJ/mol = -105 kJ/mol (-25 kcal/mol)

INSTRUCTIONS FOR PROBLEM-SOLVING

1. *Solve each problem with your peers*
2. *Compare your answers*
3. *Check your answer with an AI model of your choice (use different AI model in your team)*
   a. *Was the answer the same or different than you and your peers obtained with different AI model?*
   b. *If the same: What additional information did you receive and from which AI model?*
   c. *If the answer is different: Why is the AI solution incorrect and where is the mistake?*
4. *Did you change your answer and why?*

WORKSHOP PROBLEMS

1. Define the mechanism of a reaction.

2. A proposed mechanism is a sequence of reactions that are deduced from what source of information?

3. Given the following reaction:

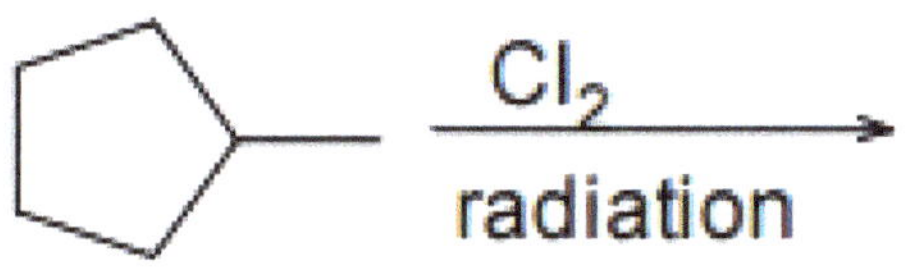

a) Predict the product(s) of the reaction.

b) Write the mechanism of the reaction.

c) Explain every step in the mechanism: molecularity, energy of activation, etc.

d) Is the mechanism you proposed concerted or multiple steps?

e) Using values from the table above calculate $\Delta H°$ for each step of the reaction and overall $\Delta H°$ for the reaction. Sketch an energy diagram for the reaction.

4. Given the following reaction:

a) Predict the product(s) of the reaction.

b) Write the mechanism of the reaction.

c) Explain every step in the mechanism: molecularity, energy of activation, etc.

d) Is the mechanism you proposed concerted or multiple steps?

e) Using values from the table above calculate $\Delta H°$ for each step of the reaction and overall $\Delta H°$ for the reaction. Sketch an energy diagram for the reaction.

5. Write the products of the following reactions, indicating major and minor products.

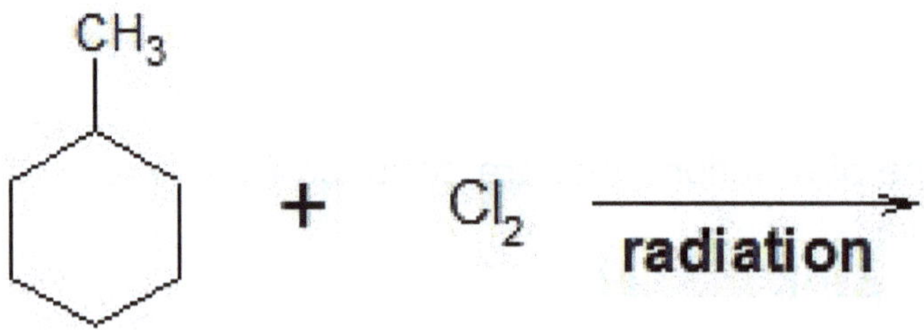

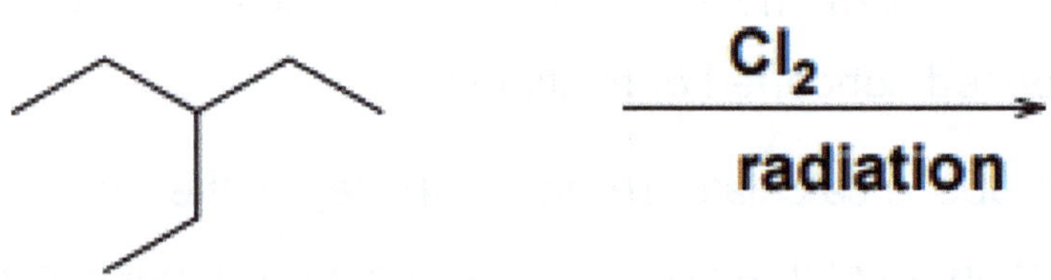

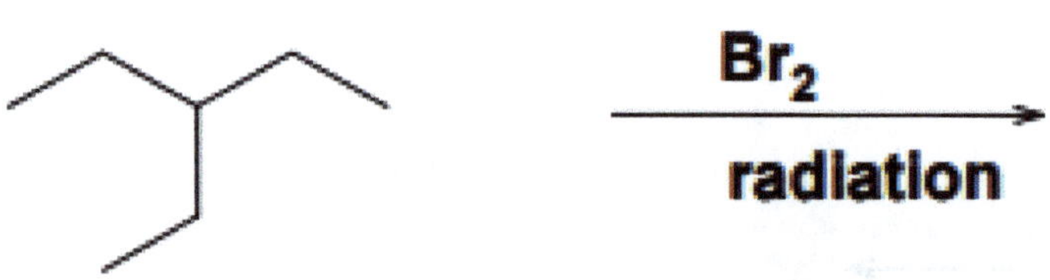

H2O

HCl

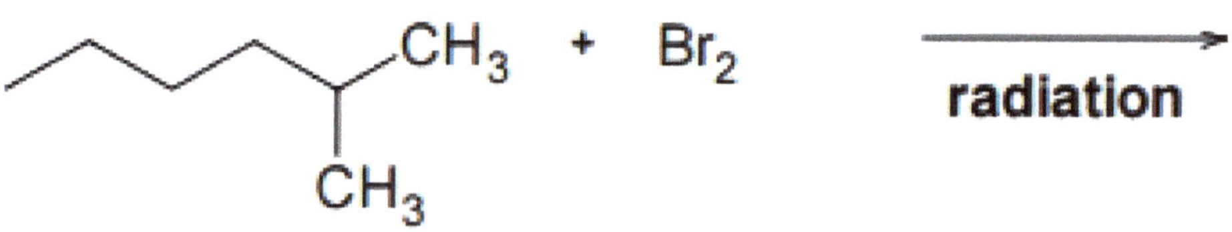
CH3
CH3
+
Br2
radiation

WORKSHOP 6 Stereochemistry

Concept Check:
- ✓ **Stereocenters**
- ✓ **Chirality**
- ✓ **Enantiomers & Diasteromers**

Chirality: property of a molecule being non-superimposable on its mirror image.

Chiral: an object that has a mirror image different from the original object.

Achiral: not chiral, identical with its mirror image.

Requirement for chiral carbon

Carbon that is bonded to four different substituents; no plane of symmetry in the overall molecule.

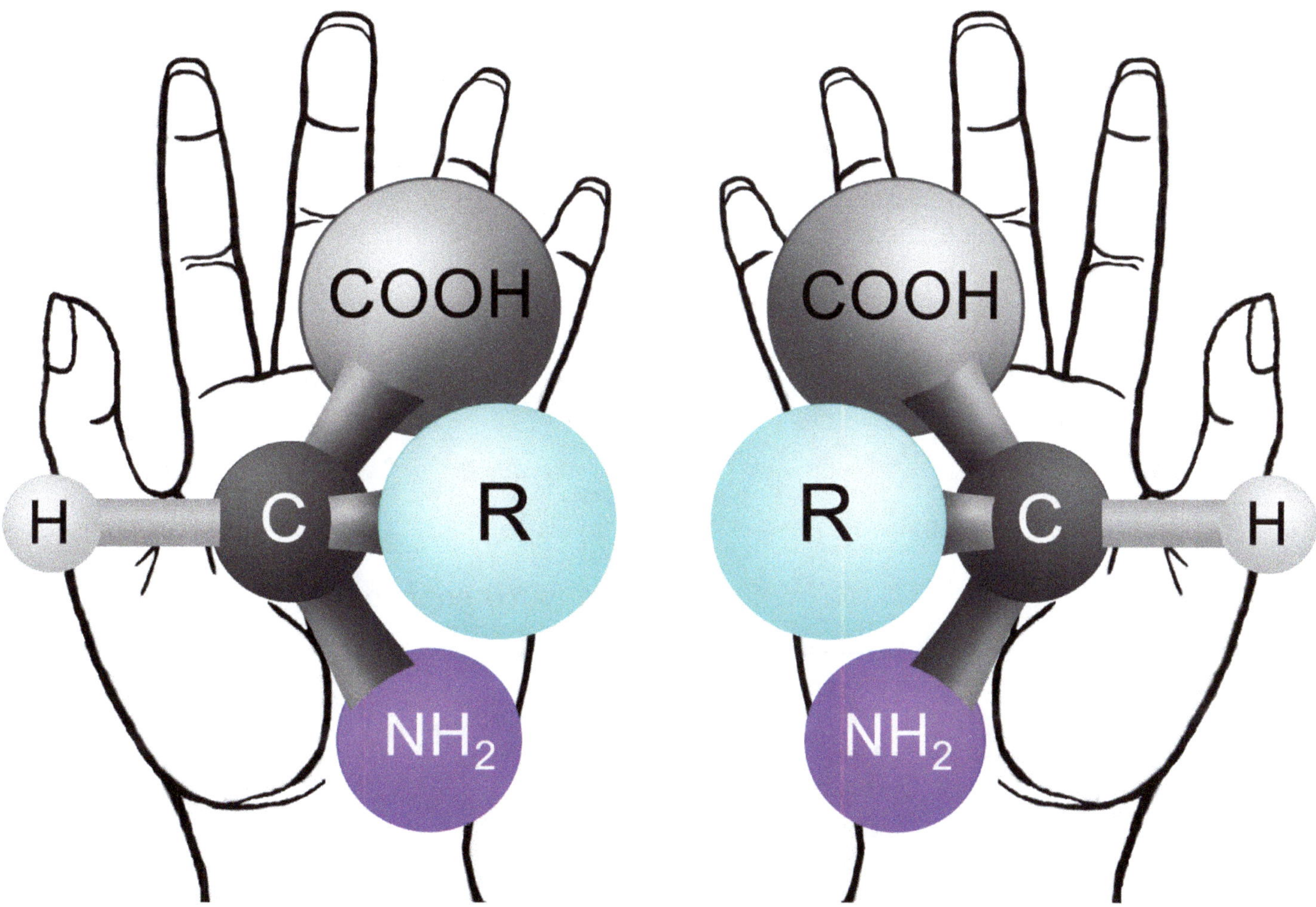

Define and discuss the following terms

Optical Activity:

Specific Rotation:

Enantiomers:

Diastereomers:

Racemic:

Meso:

Absolute configurations (R/S)

1. Assign relative priority to each group on the asymmetric carbon. Atoms with higher atomic numbers have higher priorities. Group 1 has the highest priority, group 4 has the lowest priority. In case of ties, use the next atom directly connected to the group, and double and triple bonds are treated as if each bond were connected to a separate atom.

2. View molecule as if the lowest priority group is directed away from you, into the page. Draw an arrow from the first priority group, through the second, to the third. If the arrow is going clockwise, the asymmetric carbon is called (R), if going counterclockwise, the asymmetric carbon is called (S).

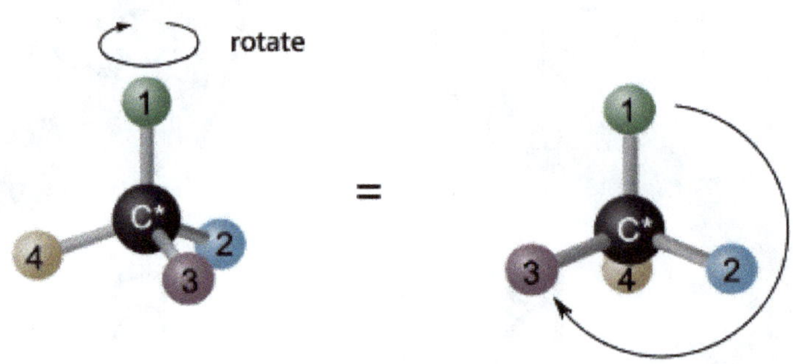

(R) enantiomer

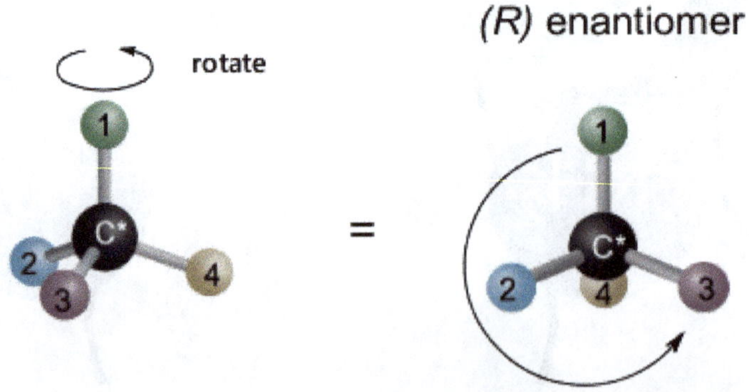

(S) enantiomer

3-D molecule to Fischer projection

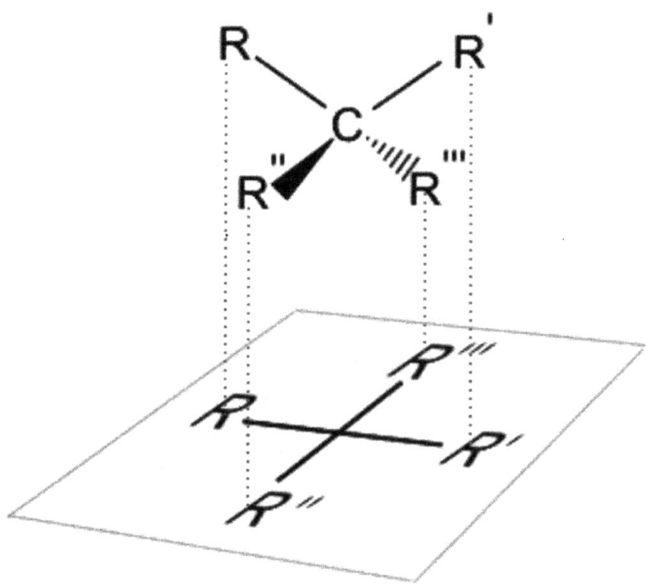

Above: Example of a Fischer projection: 3-D molecule depicted on the plane of a page in 2-dimensions. Following rules for drawing Fischer projections will allow you to consistently draw molecules in the correct stereochemical configurations.

INSTRUCTIONS FOR PROBLEM-SOLVING

1. *Solve each problem with your peers*
2. *Compare your answers*
3. *Check your answer with an AI model of your choice (use different AI model in your team)*
    a. *Was the answer the same or different than you and your peers obtained with different AI model?*
    b. *If the same: What additional information did you receive and from which AI model?*
    c. *If the answer is different: Why is the AI solution incorrect and where is the mistake?*
4. *Did you change your answer and why?*

WORKSHOP PROBLEMS

1. Identify all the stereocenters in the following compounds.

a)

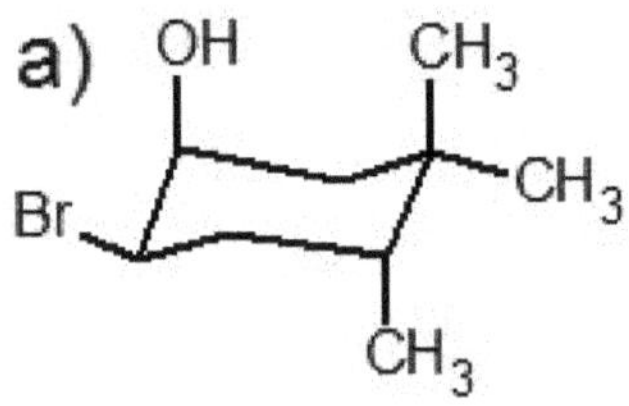

b)

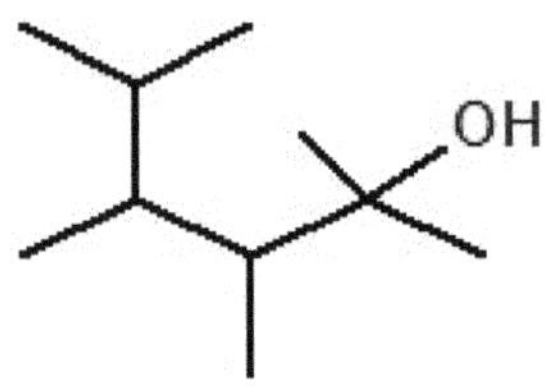

c)

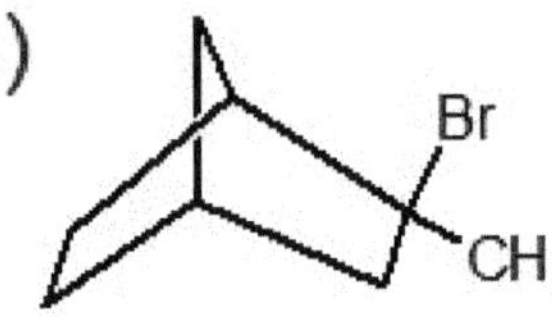

d)

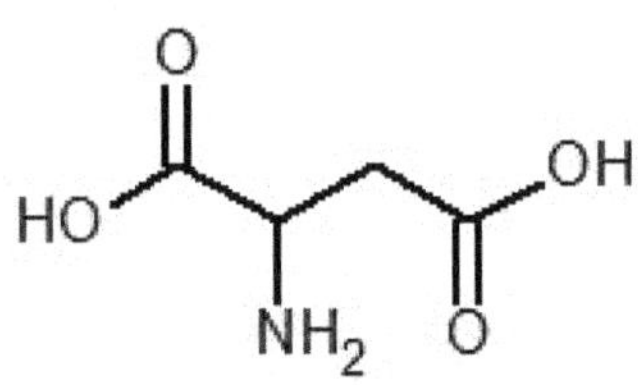

Aspartic acid

e)

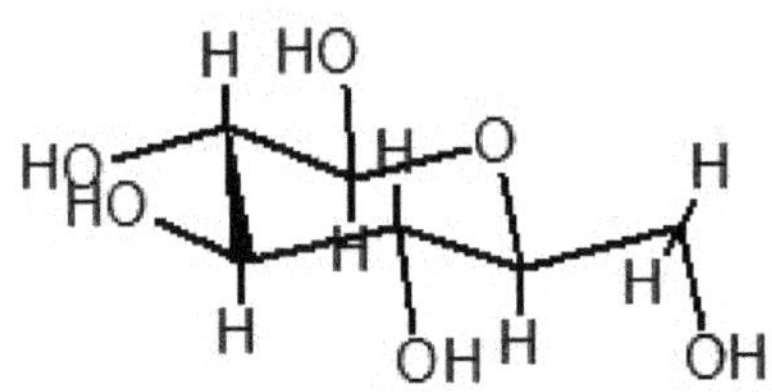

β-D-glucopyranose

f)

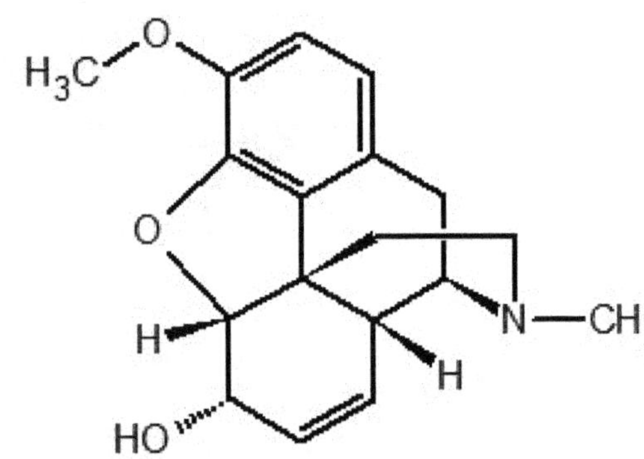

Codeine

2. Determine the relationship between any and all compounds in each of the following sets:

a) They may be identical, constitutional isomers, enantiomers, diasteromers, or not related,

b) Identify the optically active (chiral) compounds and any *meso* compounds,

c) Assign R or S configuration to all stereocenters,

d) Make molecular models to confirm your answer.

a)

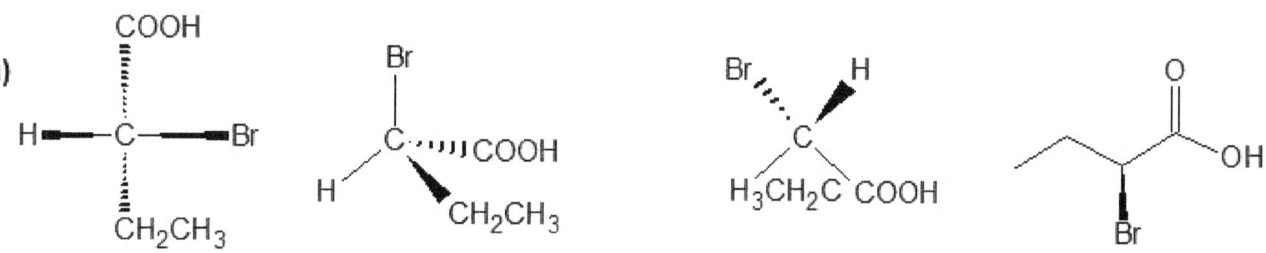

b)

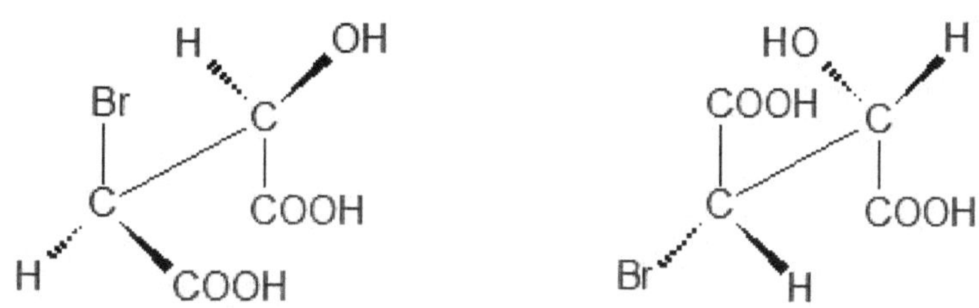

c)

d)

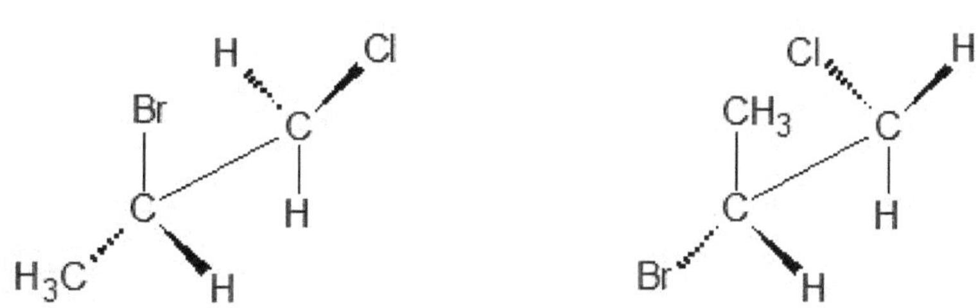

e)

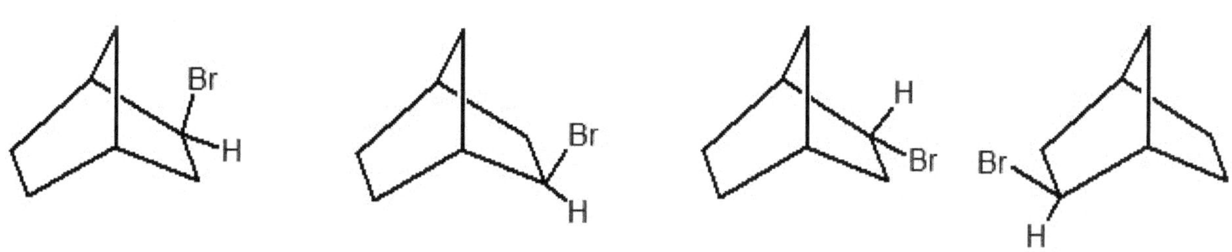

3. Determine whether the compounds in each of the following pairs are **identical, constitutional isomers, enantiomers, diastereomers,** or **not related.** Also assign a **configuration (R or S)** to all stereocenters. Identify the **optically active (chiral)** compounds. Identify any **meso** compounds.

a) 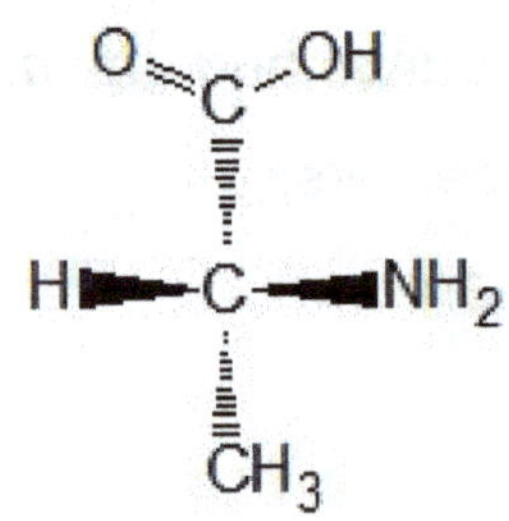 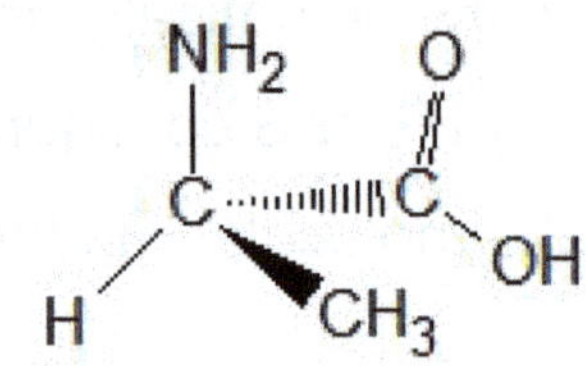

b) 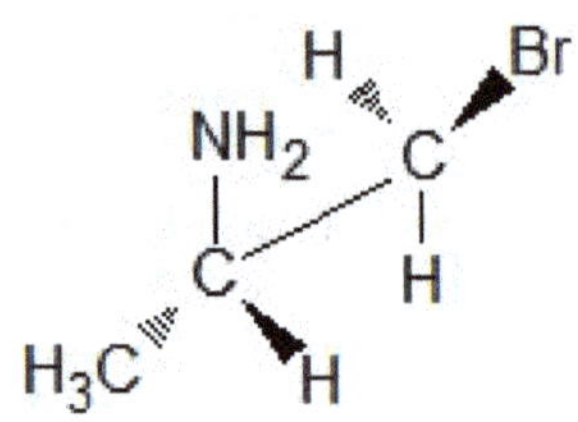 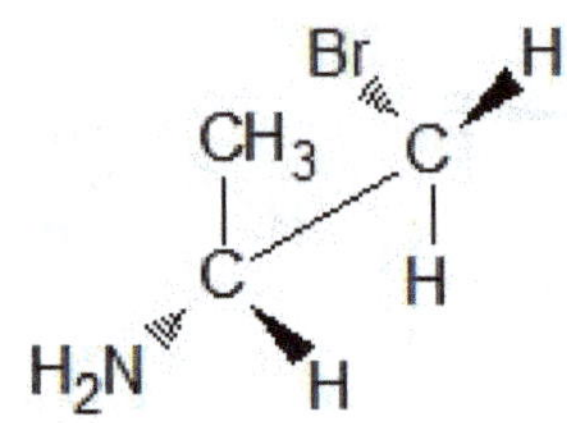

c) 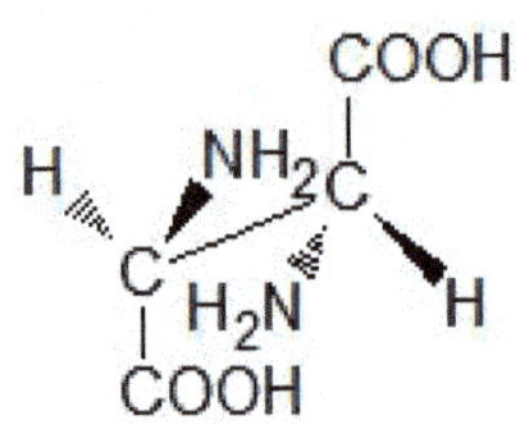 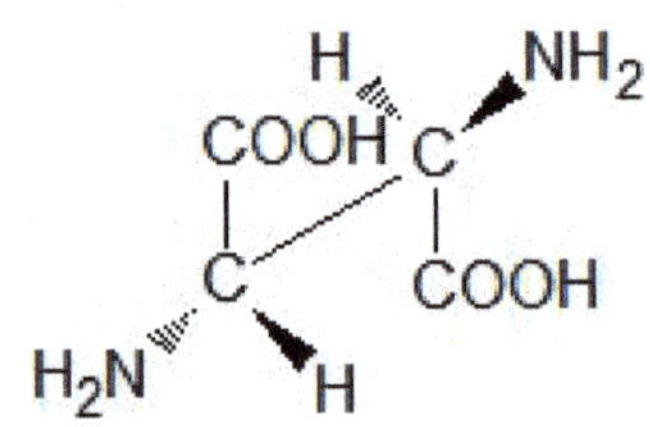

d) 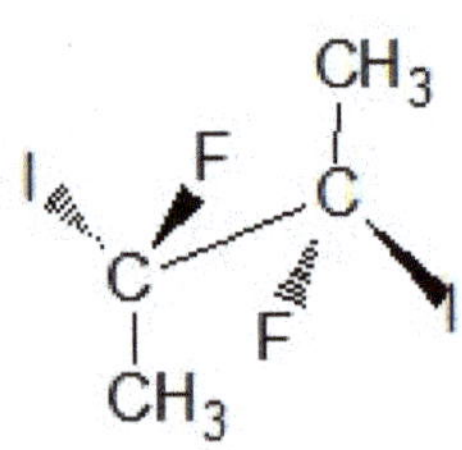 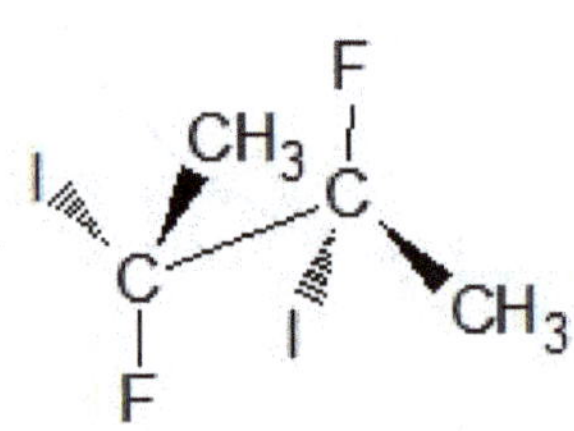

e) 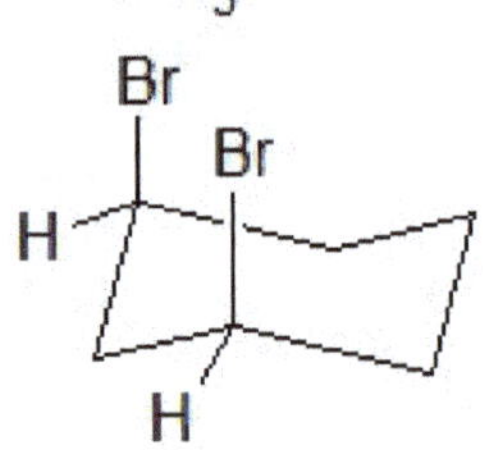 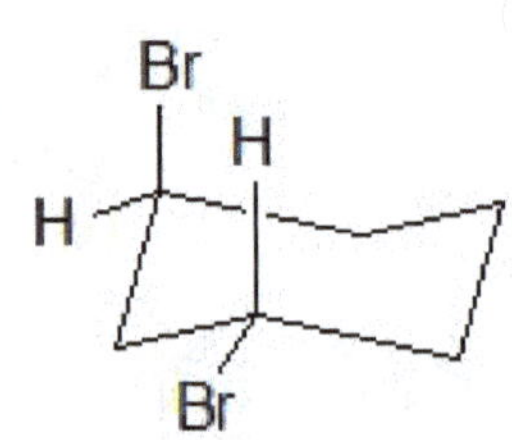

f) 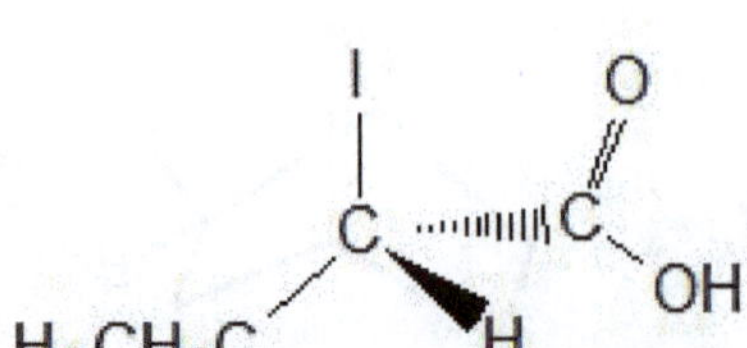 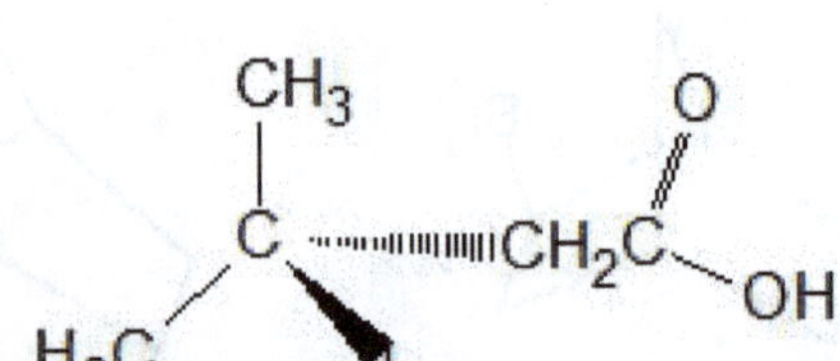

WORKSHOP 7 Nucleophilic Substitution

Concept Check:
- ✓ **Substitution reactions (SN_1 and SN_2)**
- ✓ **Nucleophiles**

Substitution Reactions

Examine the following examples of substitution reactions.
Which reactions resemble SN1 and SN2 reactions, and why? Discuss.

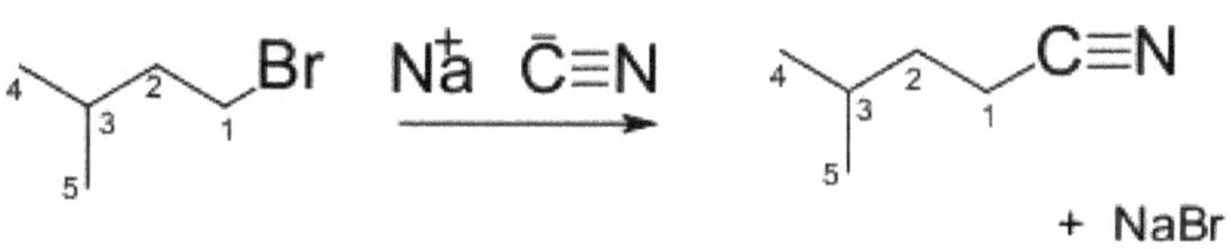

Bonds broken	Bonds formed
C_1—Br	C_1—CN
NaCN *(ionic)*	NaBr *(ionic)*

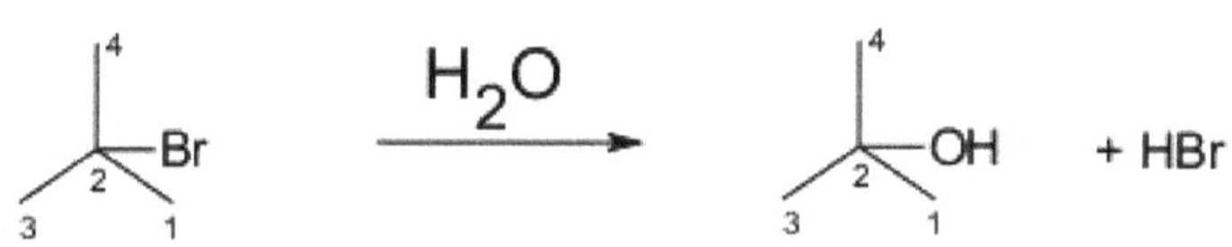

Bonds broken	Bonds formed
C_2—Br	C_2—OH
H—OH	H—Br

Trends in Nucleophilicity

1. Charge:

$^{-}SH > H_2S$ $\quad$ $^{-}NH_2 > NH_3$

A species with a negative charge is always a stronger nucleophile than its neutral counterpart. In other words, a base is more nucleophilic than its conjugated acid.

2. Decrease in nucleophilicity from left to right in the periodic table:

$^{-}NH_2 > {}^{-}OH > F^{-}$

Explain why:

____

3. Increase in nucleophilicity down the periodic table:

$I^{-} > Br^{-} > Cl^{-} > F^{-}$

Explain why:

____

INSTRUCTIONS FOR PROBLEM-SOLVING

1. *Solve each problem with your peers*
2. *Compare your answers*
3. *Check your answer with an AI model of your choice (use different AI model in your team)*
   a. *Was the answer the same or different than you and your peers obtained with different AI model?*
   b. *If the same: What additional information did you receive and from which AI model?*
   c. *If the answer is different: Why is the AI solution incorrect and where is the mistake?*
4. *Did you change your answer and why?*

WORKSHOP PROBLEMS

1. Given the following reactions, answer the following:

a) Predict the product(s) of the reaction.

b) Write the mechanism of the reaction(s).

c) Explain every step in the mechanism of the reaction: molecularity, energy of activation, etc.

d) Is the mechanism that you proposed a concerted reaction?

e) If you get more than one product, what conclusion can you reach from the mechanism of the reaction?

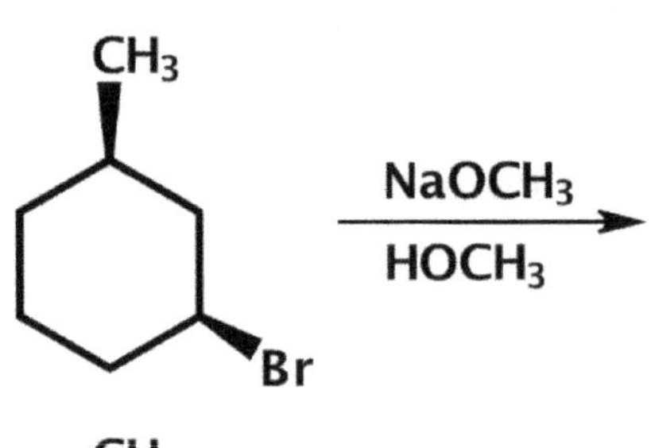

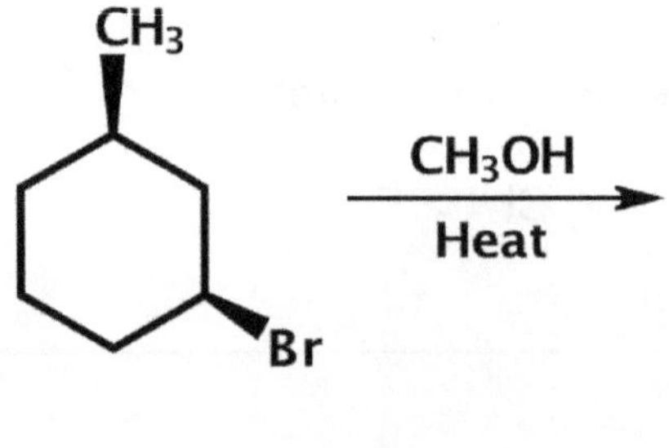

2. Select the member from each pair that shows the faster rate of SN2 reaction with H_3CO^- in acetone:

a) 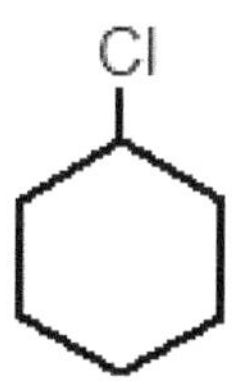 or 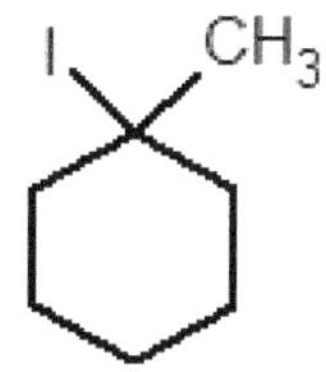

b) 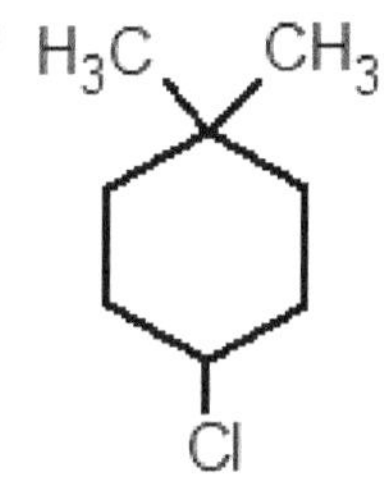 or 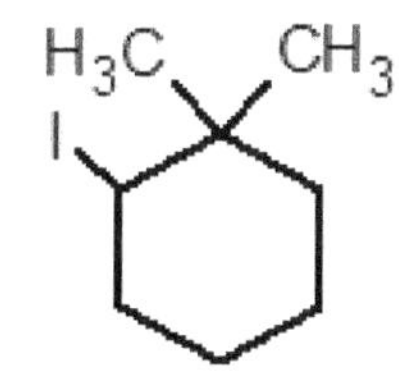

3. From each pair, select the stronger nucleophile:

a)  or 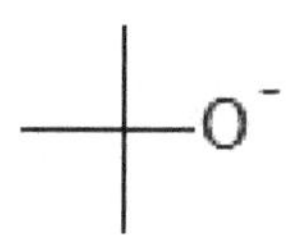

b) 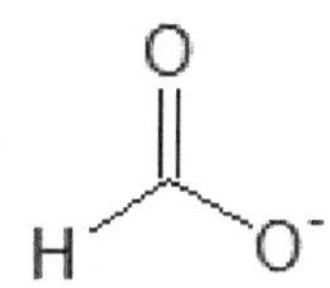 or HO^-

c) $H_3C—OH$ or $H_3C—O^-$

d)  or 

4. Draw the structural formula for the product of each SN2 reaction, including stereochemistry where appropriate:

a) 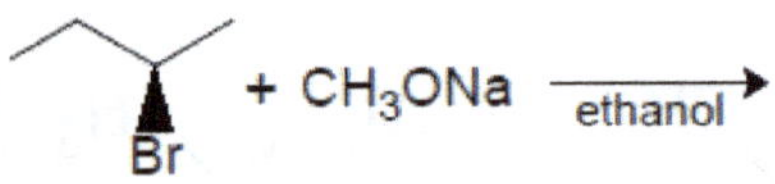

b) $(CH_3CH_2)_3N + CH_3CH_2Br \xrightarrow{\text{acetone}}$

c) 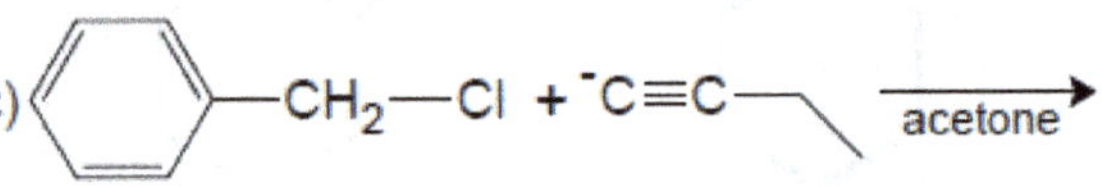

d) 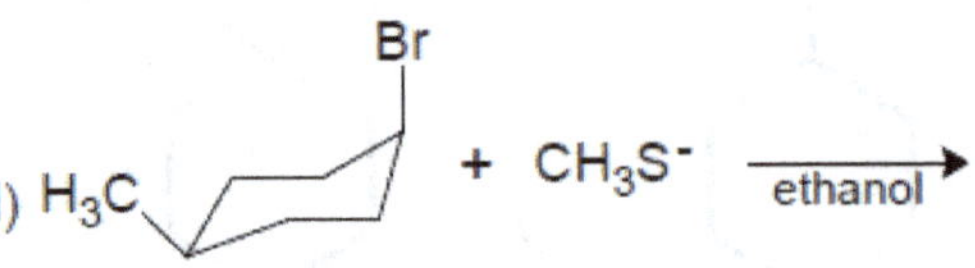

e) 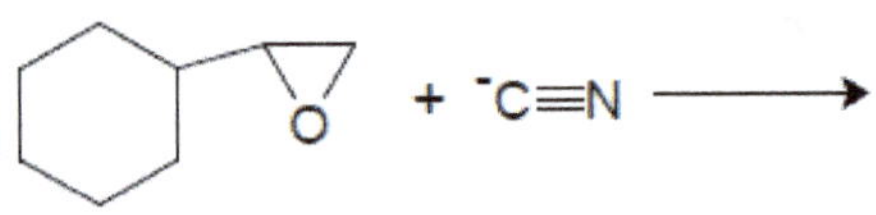

f) 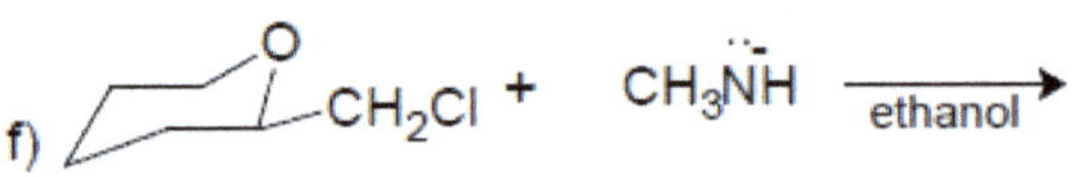

g) 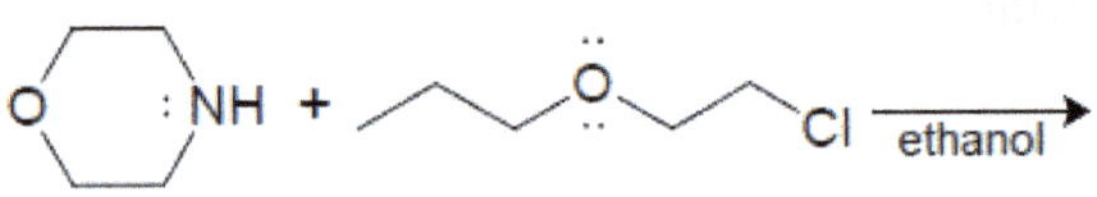

h) 

5. Select the member of each pair that undergoes SN1 solvolysis in aqueous ethanol more rapidly.

a) 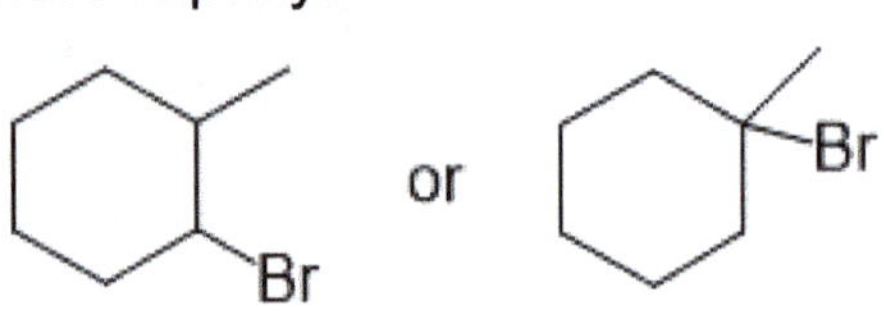

b) 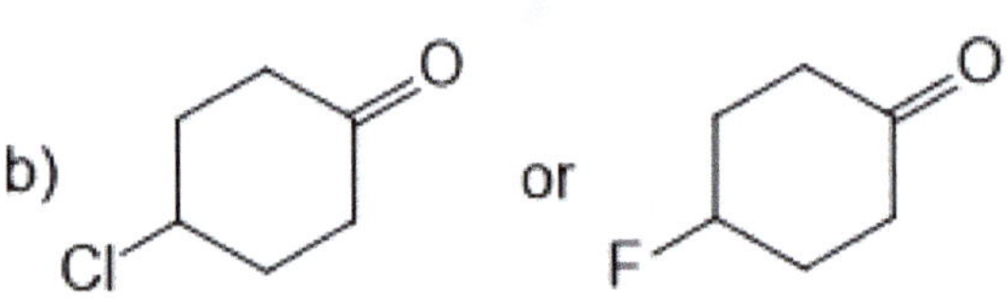

6. Draw the structural formula for the product of each SN1 reaction, including stereochemistry where appropriate:

a) 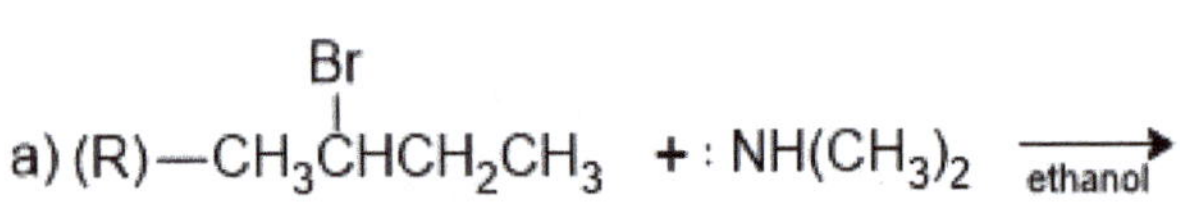

b) 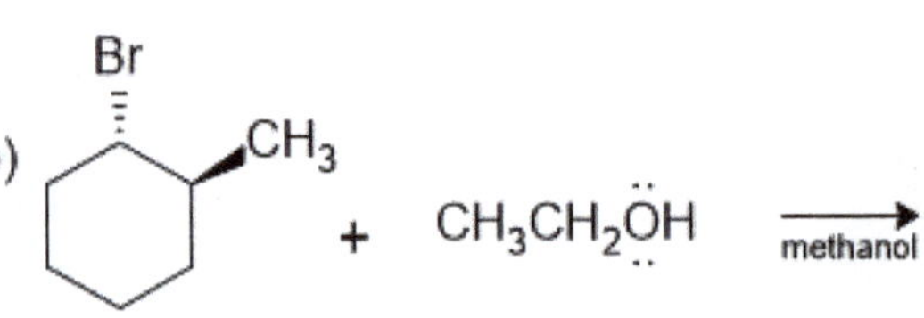

c) 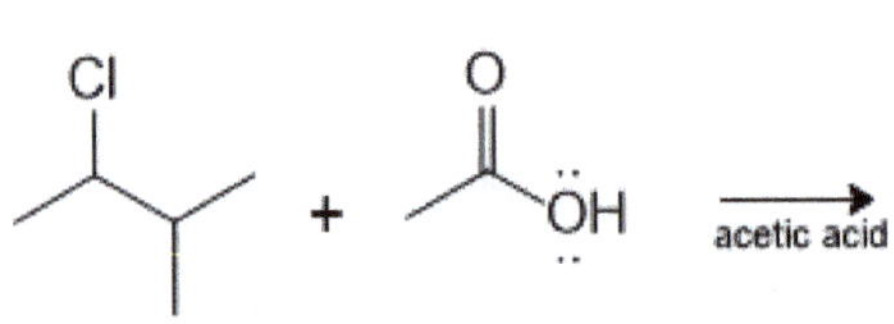

d) 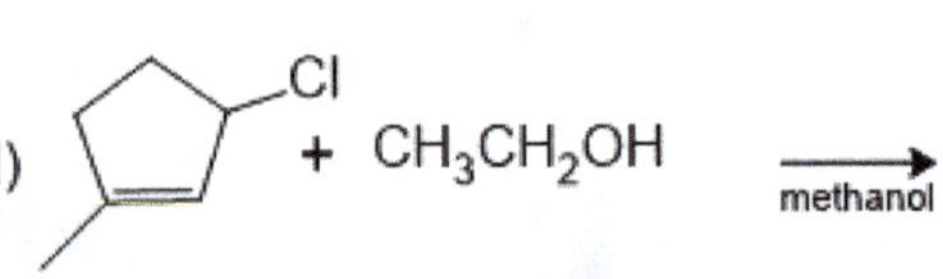

WORKSHOP 8 Substitution versus Elimination

Concept Check:
- ✓ **Elimination reactions (E_1 and E_2)**
- ✓ **Competitive types of reactions**

Elimination reactions

Elimination reactions are accompanied by a loss of two atoms or groups from the starting reactant, and usually by a formation of a new π bond. Depending on the reagents and conditions used, eliminations can be first-order or second-order.

Substitution:
1st order: Rate = *k*[substrate]
2nd order: Rate = *k*[nucleophile][substrate]

Elimination:
1st order: Rate = *k*[substrate]
2nd order: Rate = *k*[base][substrate]

Elimination reactions frequently accompany and compete with substitution reactions. By modifying the reagents and conditions used, we can promote one reaction type over the other.

E1:

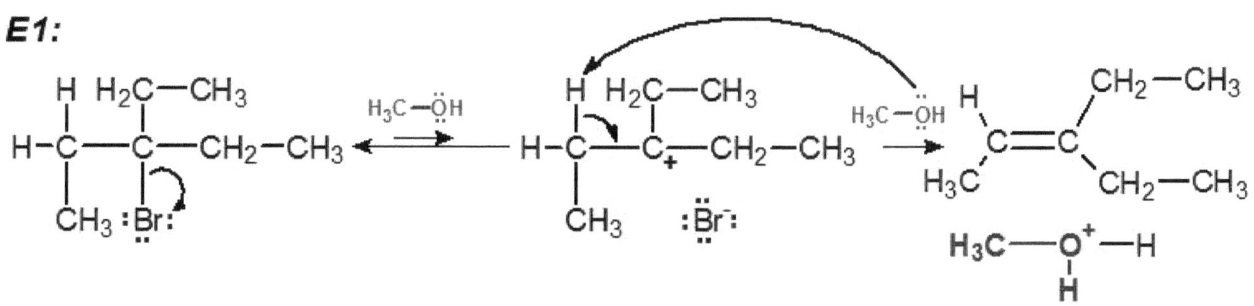

E2:

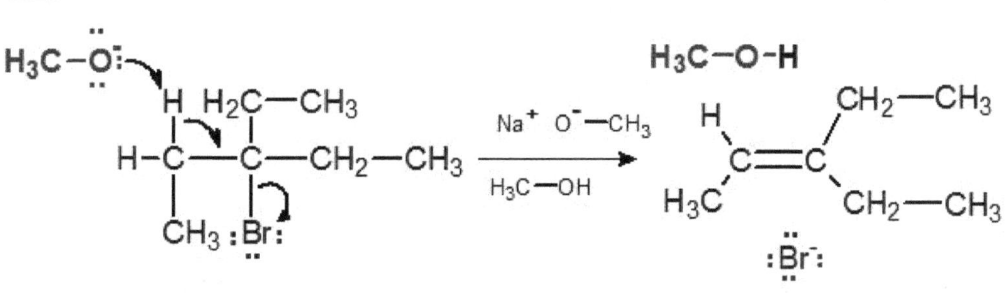

Using previous knowledge and any resources, fill in the chart and discuss:

	Carbocation (rearrangement?)	Rate dependence	Conditions	Reactivity Order 1°/2°/3°	Steps	Stereochemical requirements
S_N1						
S_N2						
E1						
E2						

INSTRUCTIONS FOR PROBLEM-SOLVING

1. *Solve each problem with your peers*
2. *Compare your answers*
3. *Check your answer with an AI model of your choice (use different AI model in your team)*
    a. *Was the answer the same or different than you and your peers obtained with different AI model?*
    b. *If the same: What additional information did you receive and from which AI model?*
    c. *If the answer is different: Why is the AI solution incorrect and where is the mistake?*
4. *Did you change your answer and why?*

WORKSHOP PROBLEMS

1. One alkyl halide, can it undergo all four reactions?
   Predict the conditions and products in all the reactions that take place

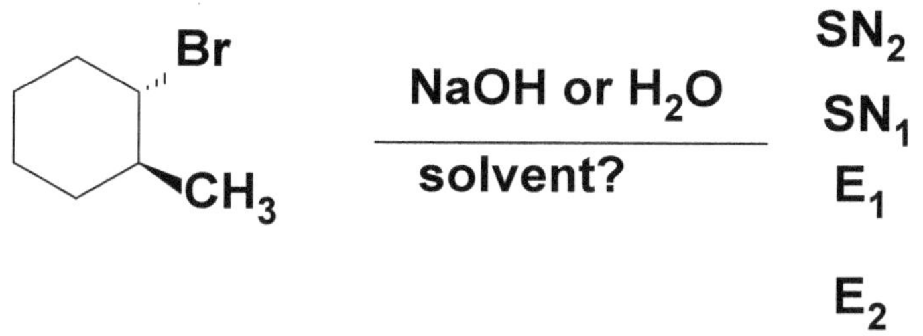

   a) Write the mechanism of the reaction for each one of them.
   b) Compare the mechanism SN1 and SN2 explain the similarities and the points of difference.
   c) Compare the mechanism E1 and E2 explain the similarities and the points of difference.
   d) Compare the four competitive reactions.

2. **a)** Draw all possible Newman projections (down the C2-C3 bond) that will allow an E2 reaction. Write the products of that reaction.

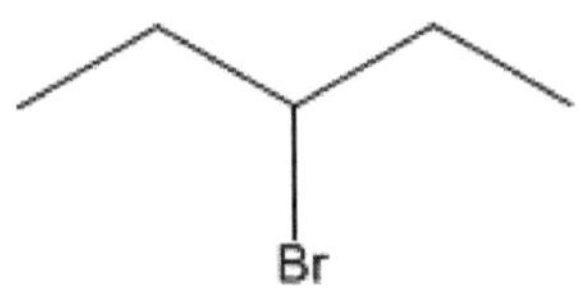

b) Do the same requirements apply to E1?

3. **a)** Rank which molecule would undergo E2 from fastest to slowest.

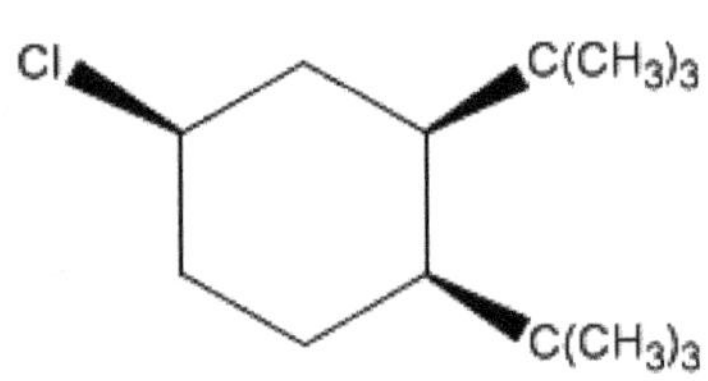

A

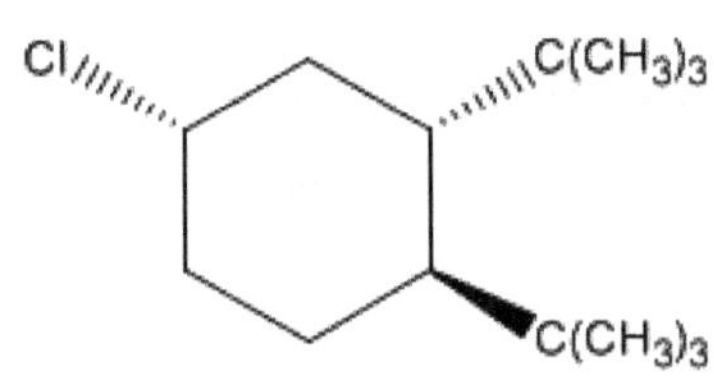

B

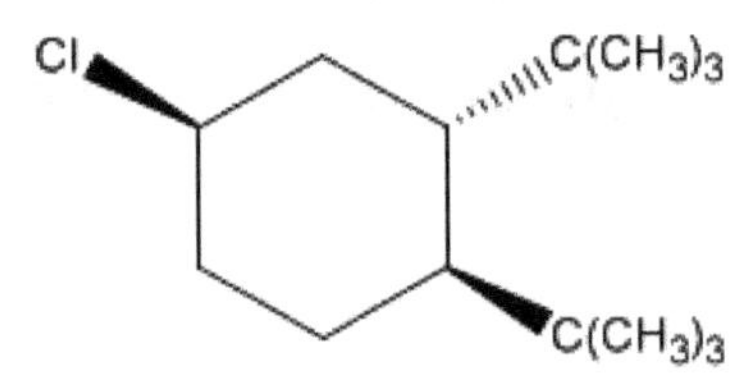

C

b) Which molecule reacts faster under conditions for E2?

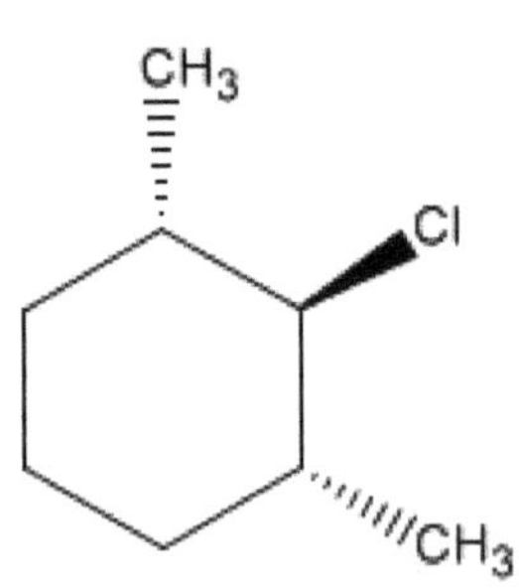

or

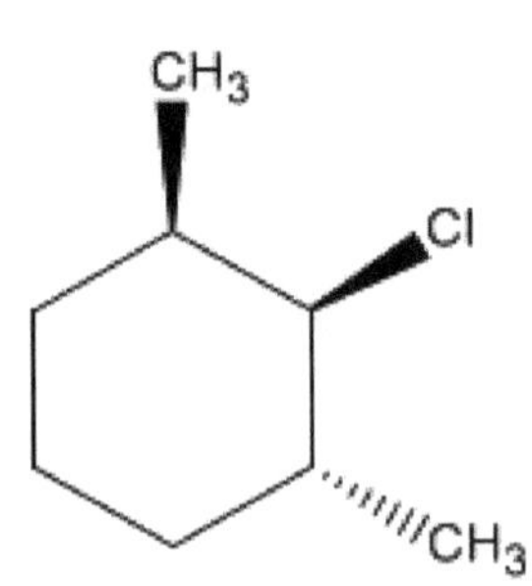

4. Predict the Products and Propose Conditions

a) Reaction progresses via the E1 mechanism:

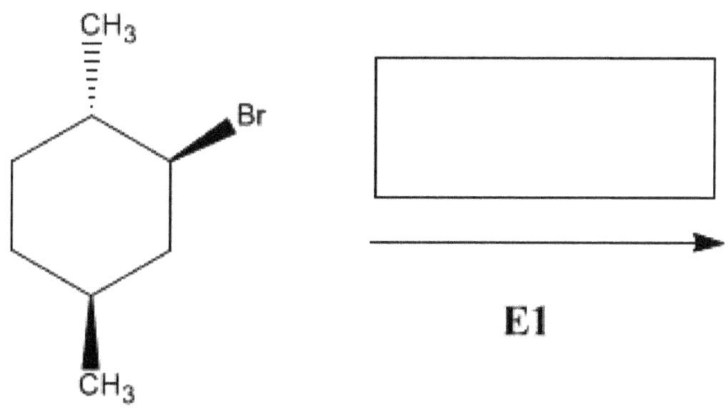

b) Reaction progresses via the E2 mechanism:

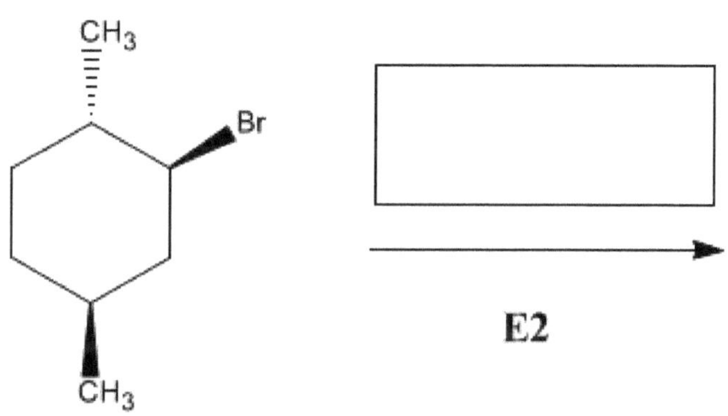

WORKSHOP 9 Alkenes

Concept Check:
- ✓ **Reactions of alkenes**
- ✓ **Elimination reactions: E1 vs. E2**

Helpful tips for visualizing molecules in mechanisms

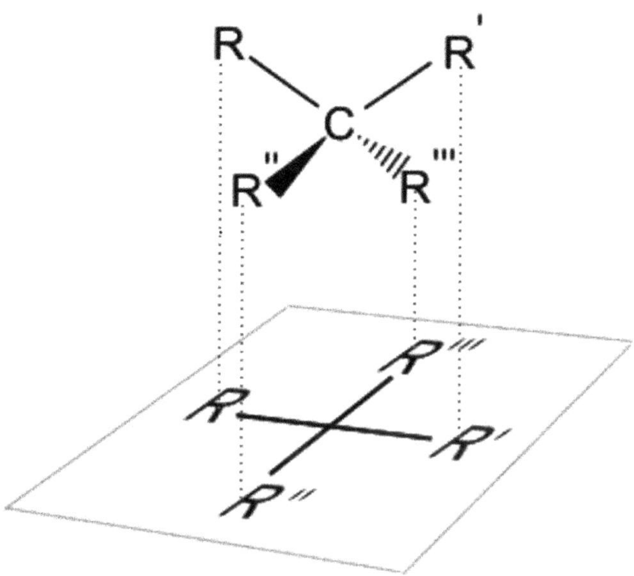

Above: Example of a Fischer projection: 3-D molecule depicted on the plane of a page in 2-dimensions. Following rules for drawing Fischer projections will allow you to consistently draw molecules in the correct stereochemical configurations.

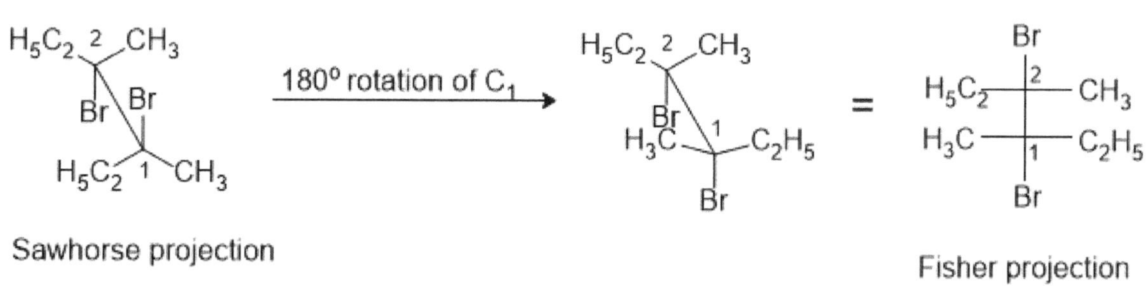

Above: Relationship of sawhorse projection to Fischer projection. Practice translating between the different ways to visualize molecules which undergo free rotation.

INSTRUCTIONS FOR PROBLEM-SOLVING

1. *Solve each problem with your peers*
2. *Compare your answers*
3. *Check your answer with an AI model of your choice (use different AI model in your team)*
    a. *Was the answer the same or different than you and your peers obtained with different AI model?*
    b. *If the same: What additional information did you receive and from which AI model?*
    c. *If the answer is different: Why is the AI solution incorrect and where is the mistake?*
4. *Did you change your answer and why?*

WORKSHOP PROBLEMS

1. Write the structures of all the isomeric alkyl bromides having the molecular formula C5H11Br, including stereochemistry where relevant.

- **a)** Which one undergoes E1 elimination at the fastest rate?
- **b)** Which one is incapable of reacting by the E2 mechanism?
- **c)** Which ones can yield only a single alkene on E2 elimination?
- **d)** For which isomer does E2 elimination gives two geometrical isomers?
- **e)** Which one yields the most complex mixture of alkenes on E2 elimination?

2. What would be the E2 elimination products conditions?

- **a)** (4R, 5S) 4-Bromo-4,5-dimethyloctane:
- **b)** (4R, 5R) 4-Bromo-4,5-dimethyloctane:

3. Given the following reactions:

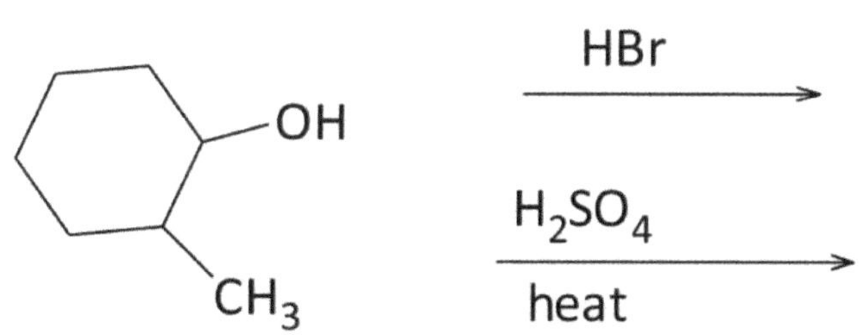

- **a)** Predict tr
- **b)** Write the mechanisms of the reactions.
- **c)** Explain each step in the mechanism of the reaction: molecularity, energy of activation etc.
- **d)** Is the mechanism that you proposed a concerted or stepwise reaction?
- **e)** If you get more than one product, what conclusion can you reach from the mechanism of the reaction?

4. Show how to carry out the following synthetic transformations using any necessary organic and inorganic reagents. More than one step may be required.

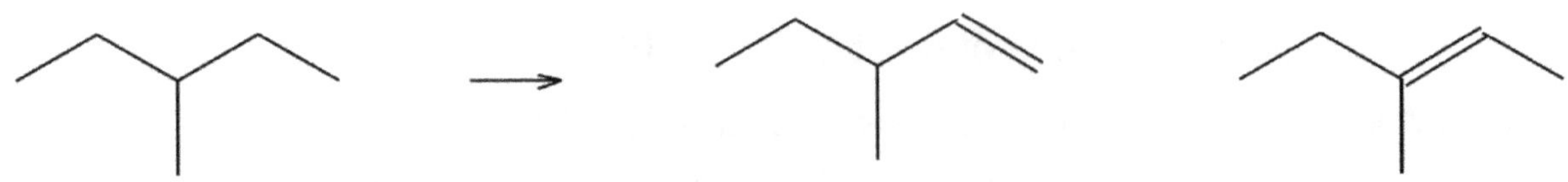

mixture of compounds predicted

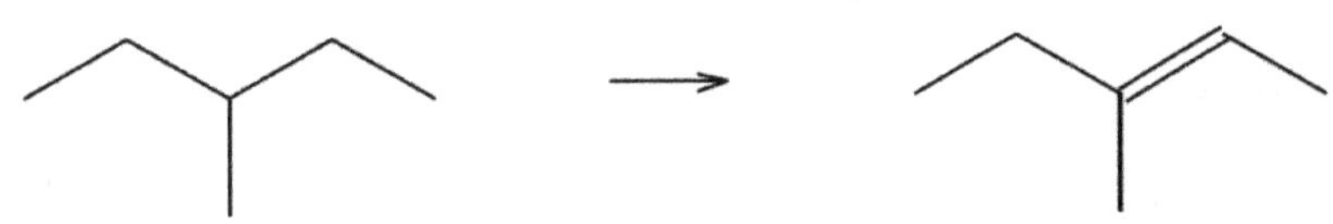

major predicted product

WORKSHOP 10 Organic Synthesis: Reactions of Alkenes

Concept Check:
- ✓ **Organic synthesis**
- ✓ **Reactions of alkenes**
- ✓ **Lengthening the carbon chain**

Guidelines for designing a synthesis

1. Look at the reaction: Identify the reactant in the product.

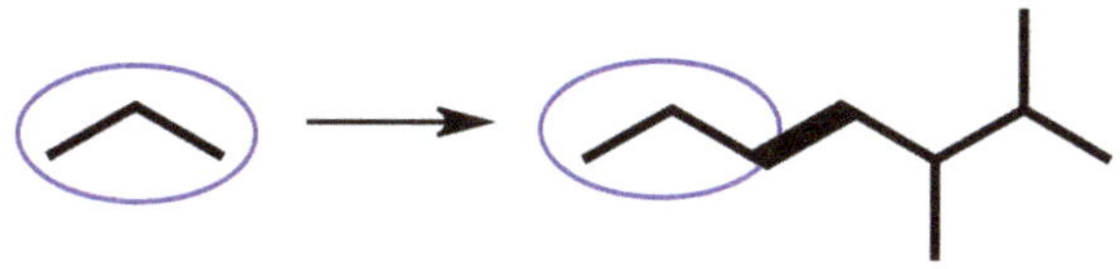

2. Find where any new bonds are formed in the product.

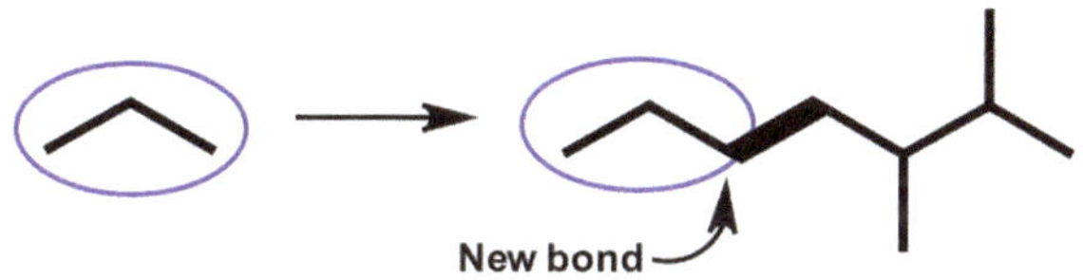

3. Determine if a functional group must be introduced to form new bonds. Functional groups characterize the reactivity of molecules. Also recognize any functional groups in the product and ask yourself, "How can I make this functional group?", or "Which molecule can I treat in one step get to this end product?"

4. This is called *retrosynthesis*, allowing you to work backwards from the product, to the intermediates, and finally to the initial reactant. The key is to visualize the needed molecule before each step, and then you can fill in the reagents afterward. Repeat these steps to reach the reactant, using appropriate reagents you have learned in class.

5. There are many ways to approach an organic synthesis. Experiment with thinking "forwards and backwards" to develop an appropriate pathway to the desired product.

Apply the synthesis guidelines given above to the sequence below. Replace the question marks with reagents or other compounds to complete the synthesis.

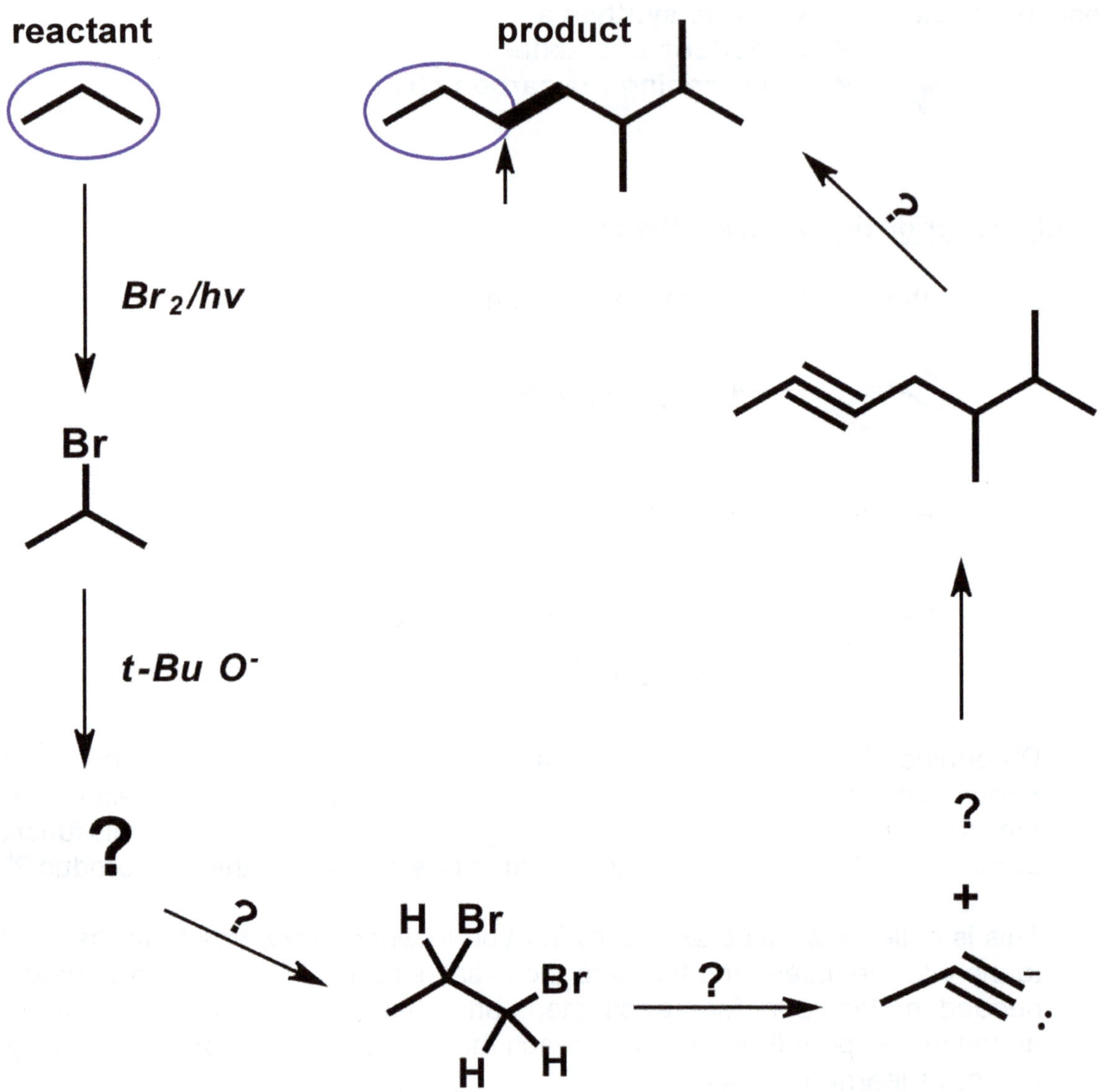

Example of a retrosynthesis pathway

Work backwards from the target product, filling in the reagent needed to get to the previous intermediate molecule. If you were writing your own retrosynthesis, you would have to figure out what molecule would get you to the product in one step, fill in the reagent, and work backwards from there to reach the reactant molecule. Use the chart on the next pages to build a reagent toolkit for all your synthesis needs.

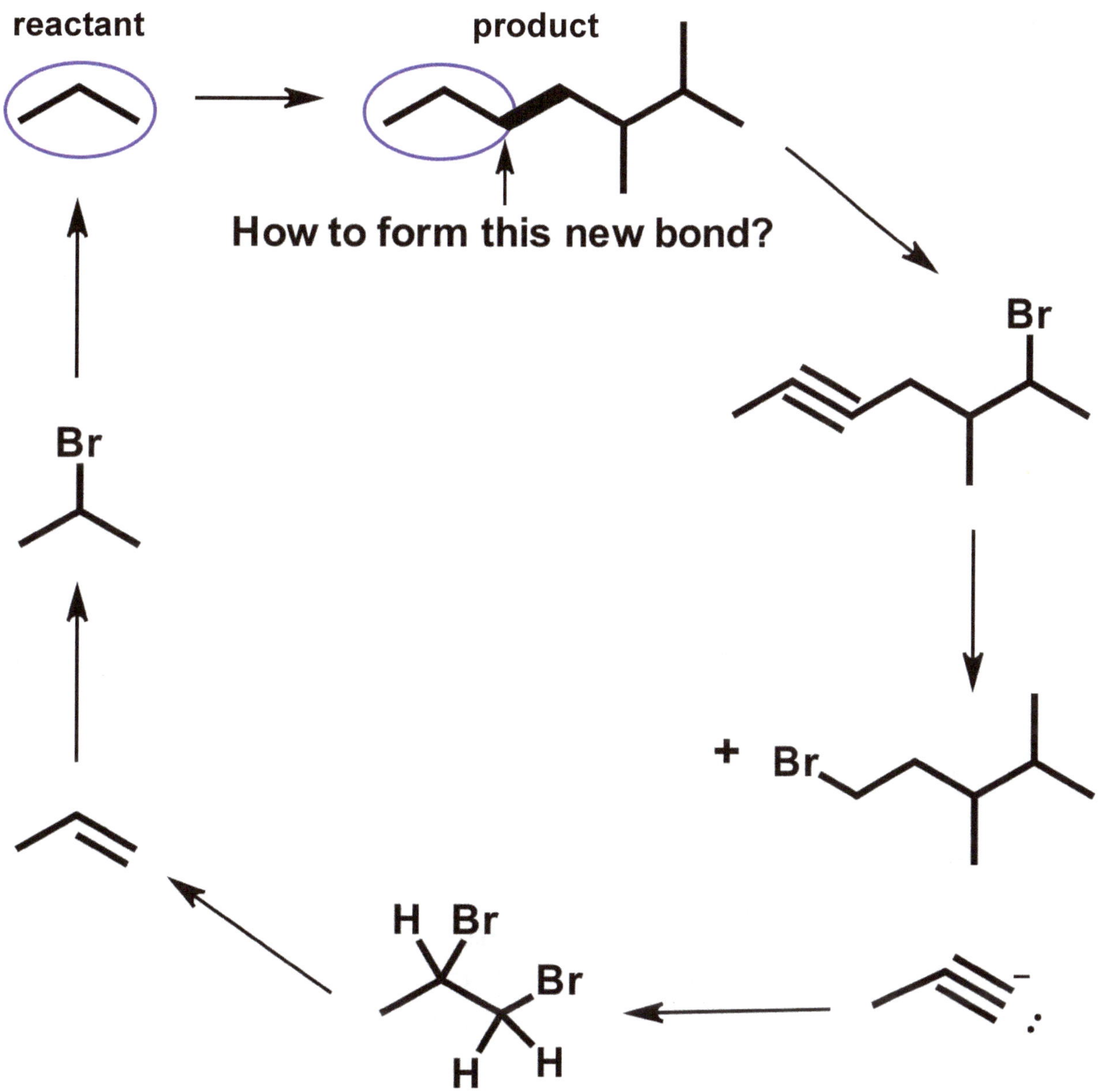

Reagent Toolkit: Fill in the missing reactants, reagents, and products to use as a reference when solving synthesis problems. A useful approach is to fill in the toolkit boxes as you work through practice problems.

Reactant	Reagent	Product	Notes
Ex.) >C=C<	HX → (HX = HCl, HBr, or HI)	–C–C– (H, X)	Markovnikov orientation (anti-Markovnikov w/HBr and peroxides)
>C=C<	→	–C–C– (H, OH)	
>C=C<	1) $Hg(OAc)_2$, H_2O → 2) $NaBH_4$		
>C=C<	→	–C–C– (RO, H)	
>C=C<	1) $BH_3 \cdot THF$ → 2) H_2O_2, ^{-}OH		
>C=C<	→	–C–C– (H, H)	syn addition
>C=C<	X_2 → (X = Cl, Br, I)		
	NBS, *hv* →		
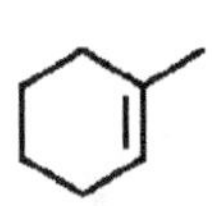	→	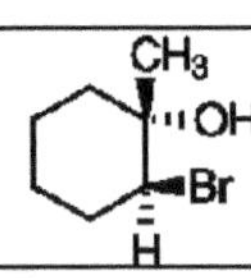	
>C=C<	+ R–C(=O)–O–O–H → (peroxyacid)		
>C=C<	→	–C–C– (OH, OH)	anti
>C=C<	cold, dilute $KMnO_4$ + ^{-}OH, H_2O (or OsO_4, H_2O_2 →		
$R_2C{=}CR'H$	→	$R_2C{=}O$ + $O{=}CR'H$ ketones & aldehydes	

Reactant	Reagent	Product	Notes
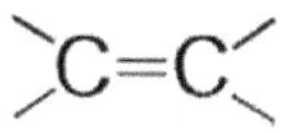	$\xrightarrow[\text{(warm)}]{KMnO_4}$		
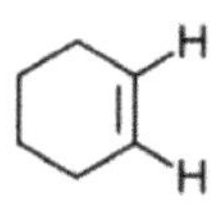	$\longrightarrow$	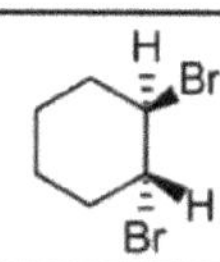	
R–C≡C–H	$\xrightarrow{NaNH_2}$		
R–C≡C–R′	$\longrightarrow$	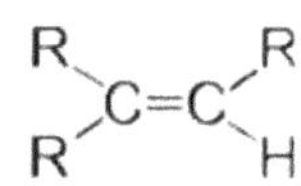	
R–C≡C–R′	$\xrightarrow{NaNH_3}$		
	$\longrightarrow$		
	$\longrightarrow$		
	$\longrightarrow$		
	$\longrightarrow$		
	$\longrightarrow$		
	$\longrightarrow$		
	$\longrightarrow$		
	$\longrightarrow$		

INSTRUCTIONS FOR PROBLEM-SOLVING

1. *Solve each problem with your peers*
2. *Compare your answers*
3. *Check your answer with an AI model of your choice (use different AI model in your team)*
   a. *Was the answer the same or different than you and your peers obtained with different AI model?*
   b. *If the same: What additional information did you receive and from which AI model?*
   c. *If the answer is different: Why is the AI solution incorrect and where is the mistake?*
4. *Did you change your answer and why?*

WORKSHOP PROBLEMS

1. Show how to carry out the following synthetic transformations using any necessary organic and inorganic reagents. More than one step may be required. Indicate with an asterisk any products formed as racemic mixtures.

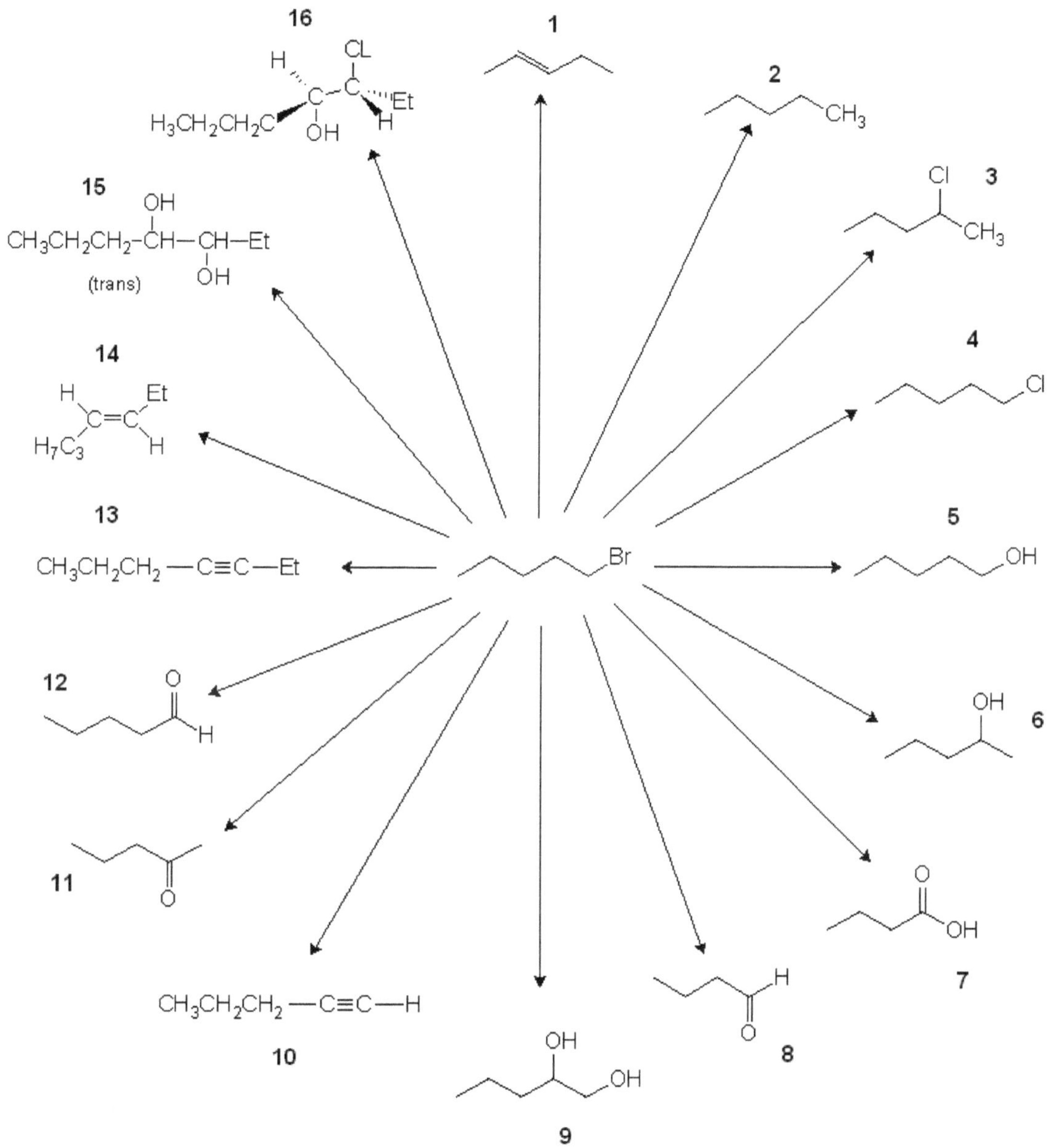

ADDITIONAL PROBLEMS Alkenes, alkynes, synthesis and reactions

1. Propose a synthesis for the following transformations:

a)

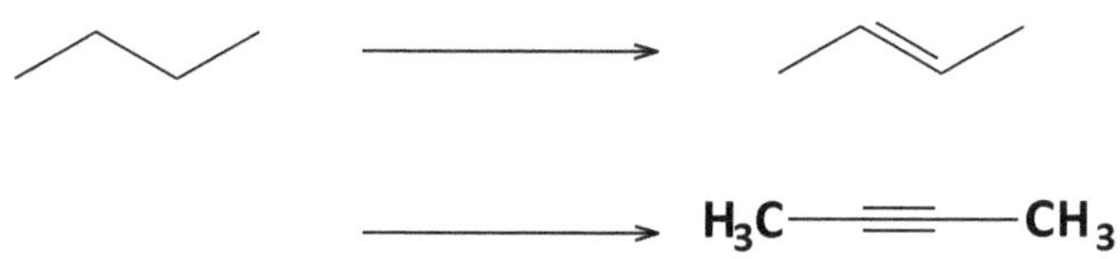

b) $CH_3 - C{\equiv}CH$

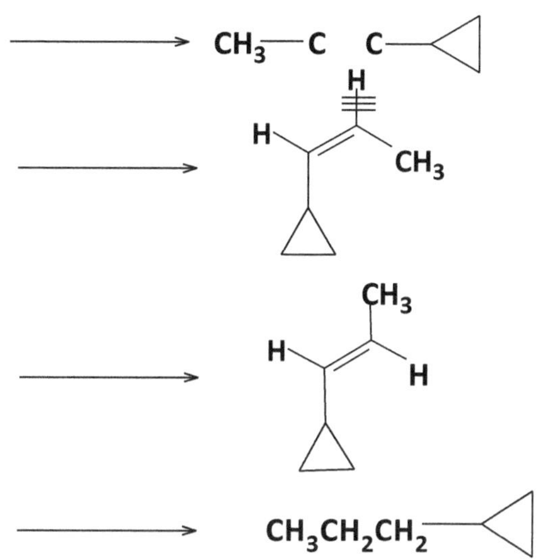

c)

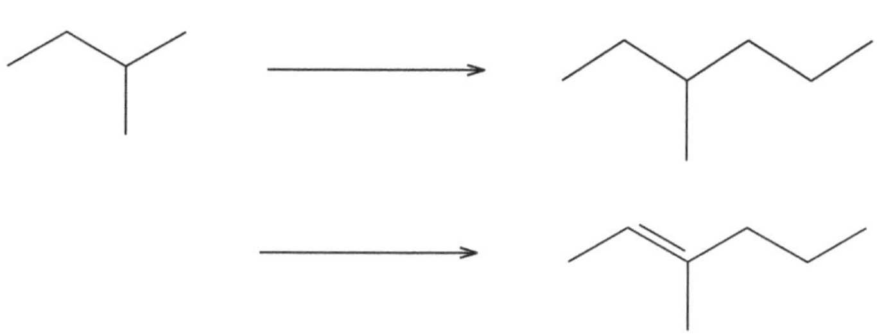

2. Write the product for the following reactions:

a)

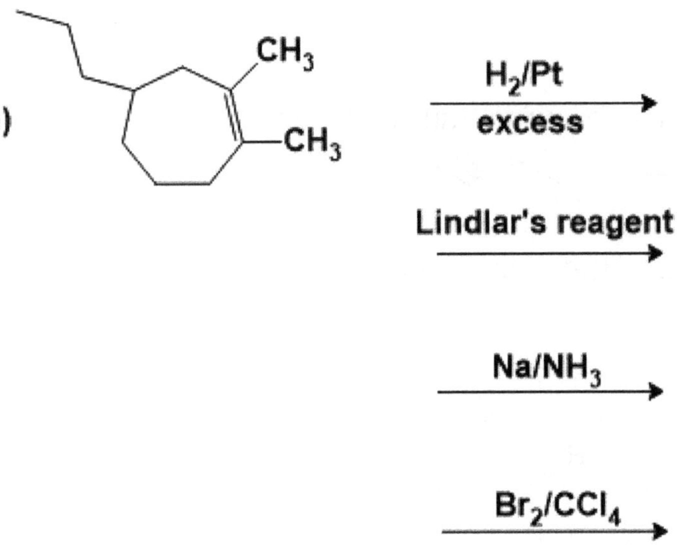

b)

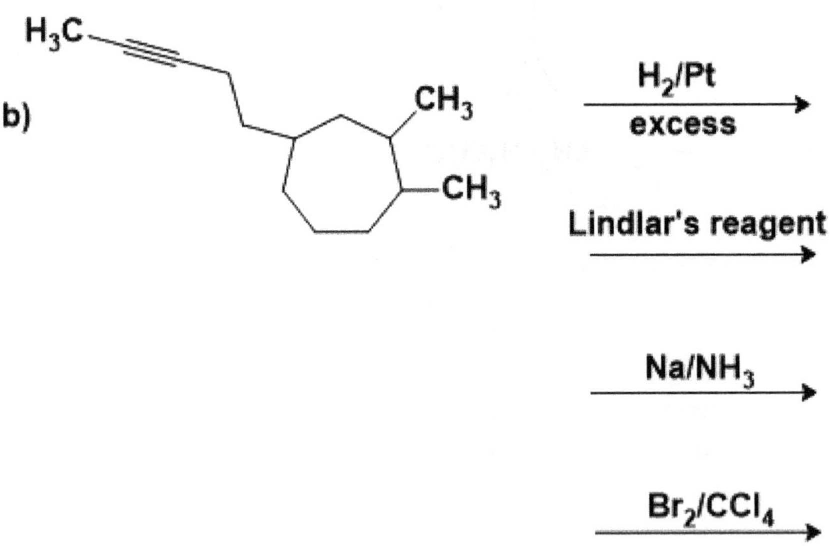

c) 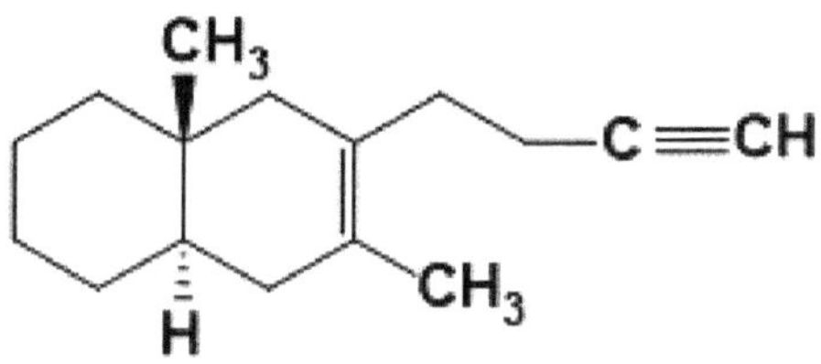

$\xrightarrow[\text{excess}]{H_2/Pt}$

$\xrightarrow{\text{Lindlar's reagent}}$

$\xrightarrow{Na/NH_3}$

$\xrightarrow{Br_2/CCl_4}$

d) 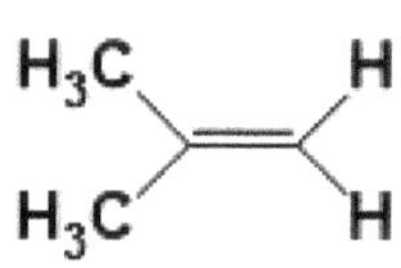

$\xrightarrow{HBr}$

$\xrightarrow[H_2O_2]{HBr}$

$\xrightarrow[OH^- \text{ cold}]{KMnO_4}$

$\xrightarrow[\text{heat}]{KMnO_4}$

$\xrightarrow[H_2O]{O_3}$

e) $H_3C-C\equiv CH$ $\xrightarrow{HBr}$

$\xrightarrow[H_2O_2]{HBr}$

$\xrightarrow[OH^- \text{ cold}]{KMnO_4}$

$\xrightarrow[\text{heat}]{KMnO_4}$

$\xrightarrow[H_2O]{O_3}$

ACID	pK_a	CONJUGATE BASE
(Strongest Acid)		*(Weakest Base)*
$HO-SO_2-OH$	-9	$HO-SO_2-O^-$
$H-I$	-9	I^-
$H-Cl$	-7	Cl^-
$CH_3CH_2OH_2^+$	-2.4	CH_3CH_2OH
H_3O^+	-1.7	H_2O
HNO_3	-1.3	NO_3^-
Cl_3CCOOH	0.64	Cl_3CCOO^-
$H-F$	3.2	F^-
C_6H_5COOH	4.2	$C_6H_5COO^-$
$C_6H_5NH_3^+$	4.6	$C_6H_5NH_2$
CH_3COOH	4.8	CH_3COO^-
$HOCOOH$	6.5	$HOCOO^-$
$CH_3COCH_2COCH_3$	9.2	$CH_3COCH^-COCH_3$
$H-C\equiv N$	9.1	$^-C\equiv N$
NH_4^+	9.4	NH_3
C_6H_5OH	10.0	$C_6H_5O^-$
^-OCOOH	10.2	$^-OCOO^-$
$^-O-N^+(=O)-CH_3$	10.2	$^-O-N^+(=O)-CH_2^-$
CH_3CH_2SH	10.5	$CH_3CH_2-S^-$
CH_3CONH_2	15.0	CH_3CONH^-
H_2O	15.7	$H-O^-$
CH_3CH_2OH	17	$CH_3CH_2-O^-$
CH_3COCH_3	19	$CH_3COCH_2^-$
Cl_3C-H	20	Cl_3C^-
$H-C\equiv C-H$	26	$H-C\equiv C^-$
H_2	35	H^-
NH_3	36	NH_2^-
$H_2C=CH_2$	36	$H_2C=CH^-$
$C_6H_5CH_3$	41	$C_6H_5CH_2^-$
C_6H_6	43	$C_6H_5^-$
CH_4	49	CH_3^-
(Weakest Acid)		*(Strongest Base)*

www.ingramcontent.com/pod-product-compliance
Lightning Source LLC
Chambersburg PA
CBHW061357240926
51846CB00007B/1442